# Back-of-the-Envelope Physics

# Back-of-the-Envelope Physics

**Clifford Swartz**

The Johns Hopkins University Press
Baltimore and London

Printed in the United States of America on acid-free paper
9 8 7 6 5 4 3 2 1

The Johns Hopkins University Press
2715 North Charles Street
Baltimore, Maryland 21218-4363
www.press.jhu.edu

Library of Congress Cataloging-in-Publication Data

Swartz, Clifford E.
Back-of-the-envelope physics / Clifford Swartz.
p. cm.
Includes bibliographical references and index.
ISBN 0-8018-7262-6 (acid-free paper) —
ISBN 0-8018-7263-4 (pbk. : acid-free paper)
1. Physics—Miscellanea. I. Title.
QC75 .S83 2003
530—dc21 2002016143

A catalog record for this book is available from the British Library.

# CONTENTS

# PREFACE

I often say that when you can measure what you are speaking about, and express it in numbers, you know something about it. But when you cannot express it in numbers, your knowledge is of a meagre and unsatisfactory kind; it may be the beginning of knowledge, but you have scarcely in your thoughts, advanced to the stage of science, whatever the matter may be.
—Lord Kelvin (William Thomson)

Well spoken, Lord Kelvin! This book is a celebration of that philosophy and of the power of the quantitative approach in physics. Quantitative is different from mathematical. Mathematical derivations are necessary for most of science, but when the math is done we must end up with numbers that can be tested or compared. Is a quantity larger or smaller than some other familiar quantity? Is our answer good to 1%, 10%, or an order of magnitude? Perhaps, for the purpose at hand, an order-of-magnitude calculation is sufficient. Perhaps the problem requires consideration of a 10% factor, but does not justify including the complicated 1% terms. Of course, at the next level of inquiry, the 1% term may involve a whole new field of physics.

In this book there are over 100 examples of interesting—and, in some cases, little-known—results of putting numbers into simple equations. With a couple of lines of arithmetic you can show that all atoms have (about) the same radius. By using the actual numerical value of a slope in a phase diagram, you can show that the pressure on an ice skate does not melt the ice. Does drinking ice water provide negative calories? Put the numbers in and find out.

Calculations that yield results good to an order of magnitude are known as *Fermi questions*. Enrico Fermi was famed for his ability to extract numerical approximations to problems for which no data seemed to exist.* In this collection of problems, however, we can feel free to look up handbook data and in most cases will calculate solutions to within a factor of two. In some cases the answers will point to further solutions or indicate the limitations of a model.

Two examples for problems of this type have run as a series of columns in the *American Journal of Physics*. One, by Victor Weisskopf, ran under the title "A Search for Simplicity" and appeared from January to December 1985. The other consisted of three questions each month, followed by solutions the following month. These were called "Back of the Envelope" problems and were presented and edited by Edward Purcell, running from January 1983 to July 1984. The level of these problems was in general higher than those presented in this book.

"Back of the Envelope" refers to the habit of many physicists, when faced with a problem, to calculate a first approximation on a piece of scrap paper—perhaps on the back of an envelope. Of course, they must have in their mental quivers the numerical values of a few frequently used constants and unit transformations. In the note to the reader, we list a number of these, such as $1.6 \times 10^{-19}\text{J} = 1\text{eV}$.

One of the approximations that we will use is the classical model of atoms—the Bohr model in which little charged spheres travel in orbits and spin on their axes. In many cases the model yields surprising agreement with experiment and provides useful heuristic connections.

---

*The archetype Fermi problem is the calculation of the number of piano tuners in New York City. There are $10^7$ people in New York and therefore $2 \times 10^6$ families. One family in ten owns a piano, so there are $2 \times 10^5$ pianos in New York. If each gets tuned once every two years, there are $10^5$ tunings per year. If a tuner can tune five pianos per day and works two hundred days a year, then the city needs one hundred piano tuners. It's unlikely that there would be fewer than ten or more than one thousand. Choose your own ratios and you will probably get the same order of magnitude (plus or minus a factor of 10).

Once the numbers and their units are established in a calculation, the arithmetic remains to be done. As a practical matter, if you have a hand calculator, use it to do the multiplication and division. Lacking such a calculator, express every value in one or two digits and its power of 10. For instance,

$$r = \frac{\hbar^2}{kq^2m} = \frac{(1.05 \times 10^{-34})^2}{(9 \times 10^9)(1.6 \times 10^{-19})^2(9 \times 10^{-31})} \Rightarrow$$

$$\frac{10^{-68}}{(10^{10})(2 \times 10^{-38})(10^{-30})} \Rightarrow 5 \times 10^{-11}.$$

A hand calculator yields $5.2 \times 10^{-11}$ for this value. Each method provides a check on the other method.

Working quantitatively requires the confidence to wheel and deal with numbers, a confidence that comes only with practice. The physics classroom is probably the only place where such practice can be acquired. Whether or not students understand Newton's laws, they should at least have witnessed the power of the quantitative approach to solving problems of all kinds. Kelvin's statement may seem pompous, but Kelvin was a very great scientist.

## NOTE TO THE READER:

# Units and Approximations

The length of Noah's ark was measured in cubits. Champagne comes in magnums. Over the years, physicists have devised a standard system of units with which to measure the universe as well as day-to-day objects. The system is known as "Systeme Internationale," or S.I.

The SI unit of energy is the joule (J), a unit that is humanly sensible. If you lift a textbook with a mass of 1 kg through a height of 10 cm you have performed 1 J of work.

$$W = mgh = (1\ \mathrm{kg})(9.8\ \mathrm{N/kg})(0.1\mathrm{m}) \approx 1\ \mathrm{J}.$$

A joule equals a newton-meter.

It takes 1 J to force 1 coulomb (C) up an electric potential hill of 1 volt (V). A joule therefore is also a coulomb-volt.

In the atomic domain, the sensible unit of energy is the electron volt, eV. Since the charge on the electron is $1.6 \times 10^{-19}$ C, 1 eV = $1.6 \times 10^{-19}$ J.

In many calculations it is convenient to use eV instead of J. For instance, the energy of a photon is h$\nu$, where h is Planck's constant and h = $6.6 \times 10^{-34}$ J $\cdot$ s. For blue light with $\lambda = 4 \times 10^{-7}$ m,

$$E_{\mathrm{photon}} = (6.6 \times 10^{-34}\ \mathrm{J \cdot s})\frac{(3 \times 10^{8}\ \mathrm{m/s})}{(4 \times 10^{-7}\ \mathrm{m})} = 5 \times 10^{-19}\ \mathrm{J} = 3\ \mathrm{eV}.$$

In eV, Planck's constant is $h = 4.1 \times 10^{-15}$ eV $\cdot$ s. We then calculate directly:

$$E_{\text{photon}} = (4.1 \times 10^{-15} \text{ eV} \cdot \text{s}) \frac{(3 \times 10^8 \text{ m/s})}{(4 \times 10^{-7} \text{ m})} = 3 \text{ eV}.$$

A dentist's X-ray has a midrange of 50,000 eV. The wavelength of such a photon is

$$\lambda = \frac{hc}{E} = \frac{(4.1 \times 10^{-15} \text{ eV} \cdot \text{s})(3 \times 10^8 \text{ m/s})}{(5 \times 10^4 \text{ eV})} = 2.5 \times 10^{-11} \text{ m}.$$

The wavelength is a quarter of an Ångstrom (Å), or a quarter of the radius of an atom.

The size significance of other quantities is often more apparent if expressed in eV. For instance, Boltzmann's constant is

$$\begin{aligned} k &= (1.38 \times 10^{-23} \text{ J/molecule} \cdot \text{K}) \\ &= (8.6 \times 10^{-5} \text{ eV/molecule} \cdot \text{K}). \end{aligned}$$

The average kinetic energy of a gas molecule at room temperature is

$$\begin{aligned} E_{\text{kin}} &= \frac{3}{2} kT = \frac{3}{2} (8.6 \times 10^{-5} \text{ eV/molecule} \cdot \text{K})(293 \text{ K}) \\ &= 3.8 \times 10^{-2} \text{ eV} = \frac{1}{26} \text{ eV}. \end{aligned}$$

The units for energy can also be the units for mass, since $E \equiv mc^2$. Indeed, the mass of the subatomic particles is usually given in eV. Here are the conversions for electron and proton:

$$\begin{aligned} E_{\text{e}} &= mc^2 = (9.1 \times 10^{-31} \text{ kg})(3 \times 10^8 \text{ m/s})^2 \\ &= 8.2 \times 10^{-14} \text{ J} = 5.1 \times 10^5 \text{ eV} \approx \frac{1}{2} \text{ MeV} \end{aligned}$$

$$\begin{aligned} E_p &= (1.67 \times 10^{-27} \text{ kg})(3 \times 10^8 \text{ m/s})^2 = 1.5 \times 10^{-10} \text{ J} \\ &= 940 \text{ MeV} \approx 1 \text{ GeV} \end{aligned}$$

$$\pi^2 = 10 \quad g = 9.8 \text{ N/kg} = 9.8 \text{ m/s}^2 = G\frac{M}{r^2} \approx 10.$$

$G$ is the universal gravitational constant $= 6.67 \times 10^{-11}\ \text{N} \cdot \text{m}^2/\text{kg}^2$.
$M$ is the mass of the Earth $= 6.0 \times 10^{24}$ kg.
$r$ is the radius of the Earth $= 6.37 \times 10^{6}$ m.

A pint's a pound the world around.
A liter's a kilogram in every land.

$v$ in m/s $\times\ 2.2 = v$ in mi/hr.

The density of air at STP is $1.3\ \text{kg/m}^3$.

Magnetic moment $\mathbf{M}_{\text{proton}} = 1.4 \times 10^{-26}\ \text{A m}^2 = 8.8 \times 10^{-8}\ \text{eV/T}$.

$1\ \text{ampere} \cdot \text{m}^2 = 10^3\ \text{gauss} \cdot \text{cm}^3 = 10^{-7}\ \text{tesla} \cdot \text{m}^3$

$= 1\ \text{joule/tesla} = 1.6 \times 10^{-19}\ \text{eV/T}$.

1 eV/particle $\approx$ 25 kcal/mole (calculated in "Latent Heat of Fusion and Vaporization" in chapter 9)

To evaluate large exponentials, use these approximations:

$$e^3 \approx 20 \qquad 2^5 = 32 \quad \Rightarrow \quad 2^{10} \approx 1000$$

$$e^{40} = e(e^3)^{13} \approx e(20)^{13} = e \times 2^{13} \times 10^{13}$$

$$= e \times 8 \times 2^{10} \times 10^{13} \approx e \times 8000 \times 10^{13} \approx 2 \times 10^{17}$$

$$e^{-40} = \frac{1}{2 \times 10^{17}} = 5 \times 10^{-18}.$$

# Back-of-the-Envelope Physics

Chapter

# 1

# Force and Pressure

## BED OF NAILS

Before making a bed of nails, every fakir should calculate the spacing of nails required for a comfortable night's sleep. Begin by testing the nails to be used. Press a nail into convenient flesh with a force meter and measure the threshold force for pain.

If the threshold for pain is $10^{-1}$ N/nail, and if the fakir has a mass of 70 kg, then

$$\frac{700\ \text{N}}{10^{-1}\ \text{N/nail}} = 7000 \text{ nails are required.}$$

If the affected area when lying on the bed of nails is $2\ \text{m} \times 30\ \text{cm} = 6 \times 10^3\ \text{cm}^2$, and if all the nails act uniformly, then the area supported by each nail is

$$\frac{6 \times 10^3\ \text{cm}^2}{7 \times 10^3\ \text{nail}} \approx 1\ \text{cm}^2/\text{nail}.$$

For comfort, the spacing should be 1 nail/cm. Try it out carefully before settling down for the night.

## SIMPLE THUMBTACK

Most practical simple machines provide an impedance match between our soft hands and some hard or awkwardly shaped device. For instance, although there is some mechanical advantage in most screwdrivers, it is also important to have a handle that fits comfortably in the hand. The same is true for doorknobs and tweezers.

Consider the thumbtack as a simple machine. The diameter of the head of a typical thumbtack is 1.3 cm. The tack comes to a sharp point with indeterminate diameter, but it is about 0.03 cm. The ratio of areas is

$$\frac{(1.3)^2}{(0.03)^2} \approx 2000.$$

The force you exert with your thumb (at the head!) is the same as the force exerted by the point. But the pressure is greater by a factor of about 2000. Of course, as soon as the point enters the material, the friction along the shaft of the tack increases. For a tack being pushed into wood, the friction would be large but necessary in order to keep the tack stuck, but to accomplish this we would need another simple machine, namely a hammer. A hammer exploits the impulse product of force and time. (See "Fatal Impulse" later in this chapter.)

## HORSEPOWER

In some English prisons long ago, inmates were required to walk on a treadmill that served as a power source to grind corn. How much power can a human provide?

A standard experiment in introductory physics classes has students measure the power they produce by running upstairs. To average out the demands on their bodies, have them run up three flights of stairs, a vertical distance of about 30 feet. (The height of school floors may vary.) A reasonable running time for a 65 kg young person might be 10 s.

$$\text{Power} = \frac{\text{work}}{\text{time}} = \frac{(65\ \text{kg})(10\ \text{N/kg})(10\ \text{m})}{10\ \text{s}} = 650\ \text{W}.$$

Since 1 horsepower = 746 W, our hypothetical student was able for a short time to have an output of almost 1 hp. This is a popular exercise for students, but few can exert one horsepower for more than a few seconds. On the other hand, no horse could run up three flights of stairs.

Chinning is another popular exercise in strength tests. For a full chin-up, the body must rise about 55 cm. The work done in a chin-up is

$$W = (65\ \text{kg})(10\ \text{N/kg})(0.55\ \text{m}) = 360\ \text{J}.$$

It would take about one horsepower to do two chin-ups per second, an unlikely feat.

## BUOYANCY IN AIR

Does your bathroom scale lie because you are being buoyed up by the air? Since most people can just about float in fresh water, the density of a human body must be about the same as that of water—1000 kg/m$^3$. A human weighing 220 lb has a mass of 100 kg, and therefore a volume of 0.1 m$^3$. The density of air is 1.3 kg/m$^3$, so a human is buoyed up by the weight of 0.13 kg, which is a force of 1.3 N, or 0.29 lb. The weight that this heavy human reads on a bathroom scale is about one-third pound lighter than his actual weight.

The buoyancy on an object in air is $\rho_a\, g\, V$, where $\rho_a$ is the density of air, and $V$ is the volume of the object. The relative effect is

$$\frac{\rho_a\, g\, V}{\rho_o\, g\, V} = \frac{\rho_a}{\rho_o},$$

where $\rho_o$ is the density of the object.

For brass, such as the weights on a balance, $\rho_o = 9000\,\text{kg/m}^3$, and the relative buoyancy effect is 0.01%. Some chemical balances are so sensitive that the air buoyancy must be taken into account.

## HOW DENSE IS THE OCEAN?

In a first approximation to find the pressure in the atmosphere (see *Height of Atmosphere*), we assumed that the density of the air was a constant. That certainly isn't true, but the assumption yielded the right order of magnitude for the height of the atmosphere.

How about the density of seawater in the ocean? Let's find the pressure at great depths, and then calculate the density. Assuming constant density, the pressure is

$$P = P_0 + (\text{weight density}) \times (\text{depth}).$$

$P_0$ is atmospheric pressure, which is insignificant. (In terms of equivalent "pressure head," it is only 10 m.)

The maximum recorded ocean depth is about 10 km in the Marianas Trench, east of the Philippines. The pressure at that depth is

$$P = (1 \times 10^4\ \text{N/m}^3)(1 \times 10^4\ \text{m}) = 1 \times 10^8\ \text{N/m}^2 = 1000\ \text{atm}.$$

To find the change in density of water at this pressure, use the bulk modulus of water, which is

$$B = \frac{-\Delta P}{(\Delta V/V)} \quad \Rightarrow \quad (2 \times 10^9\ \text{N/m}^2) = \frac{-(1 \times 10^8\ \text{N/m}^2)}{(\Delta V/V)}.$$

The minus sign accounts for the fact that as the pressure increases, the volume decreases.

The fractional change in volume is equal to the fractional change in density.

$$\frac{\Delta V}{V} = \frac{\Delta \rho}{\rho} = \frac{(1 \times 10^8\ \mathrm{N/m^2})}{(2 \times 10^9\ \mathrm{N/m^2})} = 0.05.$$

At that extreme depth, the water density has increased by only 5%. For most purposes, our assumption of constant density in calculating pressure in water is good enough. To look for greater accuracy would require taking into account other factors, such as salinity and temperature.

## WASHBOWLS AND CORIOLIS FORCE

Legend has it that washbowls in the Northern Hemisphere drain counterclockwise and in the Southern Hemisphere clockwise. This behavior is supposed to be due to the Coriolis effect, which does indeed cause atmospheric hurricanes to turn counterclockwise in the Northern Hemisphere,

To find out if washbowls act the same as hurricanes, put in the numbers. Coriolis force is

$$F_{\mathrm{C}} = -2m\, \vec{\omega} \times \vec{v}.$$

The rotational velocity, $\omega$, of the rotating reference frame (the Earth) is

$$\frac{2\pi \text{ radians}}{24 \text{ hr} \times 3600 \text{ s/hr}} = 7.3 \times 10^{-5} \text{ rad/s}.$$

A reasonable radial velocity for draining water in a washbowl is 1 m/s. For the component of $\vec{\omega}$ perpendicular to the draining water, assume a factor of $^1\!/_2$. Now compare the Coriolis acceleration with the gravitational acceleration.

$$\frac{2\,\vec{\omega} \times \vec{v}}{g} = \frac{7 \times 10^{-5} \times 1}{10} = 7 \times 10^{-6} \approx 10^{-5}.$$

The Coriolis acceleration is smaller than the gravitational acceleration by a factor of about $10^5$. Experiments done at M.I.T. confirm that the legend is wrong, except in the case of very elaborate procedures that control the symmetry of the bowl and the drainage and the establishment of noncirculating water before the drainage begins. The precision experiment in which the Coriolis force was detected was described by Ascher Shapiro in the December 15, 1962, edition of *Nature*, volume 196, page 1080.

## POINTS OF CONTACT

In many cases, to a first approximation, dry sliding friction is independent of the surface area involved. That's strange! Dry friction is caused by breaking the small welds between surface features, and by plowing through the uneven surface. Why shouldn't friction depend on the number of these contact points, and thus the surface area? In fact, it does, but very few surface points are actually in contact—as we shall see.

Consider a 1 kg steel block sliding on steel. Even polished steel is pitted and mountainous on the atomic level. The force of repulsion at points in contact is $5 \times 10^{-9}$ N/atom (see *Binding Force of Atoms*). The force per atom times the number of atoms in contact must equal the weight of the 1 kg block that is being supported.

$$(5 \times 10^{-9}\ \text{N/atom}) \times \#\ \text{of atoms} = 10\ \text{N}$$

$$\#\ \text{of atoms} = 2 \times 10^{9}.$$

Each atom with a diameter of $2 \times 10^{-10}$ m covers an area of $4 \times 10^{-20}$ m$^2$. The area covered by the supporting atoms is

$$(2 \times 10^{9}\ \text{atoms}) \times (4 \times 10^{-20}\ \text{m}^2/\text{atom})$$

$$= 8 \times 10^{-11}\ \text{m}^2 = 8 \times 10^{-7}\ \text{cm}^2.$$

The 1 kg steel block has a volume of $(1000 \text{ g})(\frac{1 \text{ cm}^3}{8 \text{ g}}) = 125 \text{ cm}^3$. If the block is cubic, it will be 5 cm on each side and have a face area of 25 $\text{cm}^2$. However, only $8 \times 10^{-7} \text{ cm}^2$ is in actual contact.

$$\frac{\text{Actual area}}{\text{Apparent area}} = \frac{8 \times 10^{-7} \text{ cm}^2}{25 \text{ cm}^2} = 3 \times 10^{-8}.$$

Even if the block were 25 cm × 5 cm × 1 cm and placed so that the apparent area of contact were 125 $\text{cm}^2$, or 5 $\text{cm}^2$, the actual contact area would still be $8 \times 10^{-7} \text{ cm}^2$.

## ATMOSPHERIC FORCE

Everyone likes to see things squash. A standard physics demonstration is to create a vacuum in a rectangular metal can. As the air pressure in the metal can is reduced, the can collapses. The friendly atmosphere in which we are all submerged can create enormous unbalanced forces on surfaces that have a vacuum on one side. The atmospheric pressure is only $14.7 \text{ lb/in}^2 = 1 \times 10^5 \text{ N/m}^2$, but the total force exerted is the pressure times the area.

A typical one-gallon can has a height of 24 cm and width of 15 cm. The area of one side is $(24 \text{ cm}) \times (15 \text{ cm}) = 360 \text{ cm}^2 = 3.6 \times 10^{-2} \text{ m}^2$. The force exerted on that side is $(1 \times 10^5 \text{ N/m}^2)(3.6 \times 10^{-2} \text{ m}^2) = 3.6 \times 10^3 \text{ N}$.

Let's convert that value into more familiar units. A person with a mass of 100 kg weighs 980 N, which is only about one-fourth of the atmospheric force on the can. That person is heavy but could stand on the gallon can without collapsing it. Another way to view the magnitude of the atmospheric force on the can is this: 1000 kg weighs 1 metric tonne or 2200 lb or $9.8 \times 10^3$ N. The atmospheric force on the can is

$$\frac{3.6 \times 10^3}{9.8 \times 10^3} (1 \text{ tonne}) \approx 0.4 \text{ tons}.$$

Air pressure can exert much larger forces. It can (and does) support cars and heavy trucks (see, the next section, "Weighing Your Car"). In some particle accelerator laboratories with huge beam-deflecting magnets that must be positioned very precisely, the magnets are floated into place. The bottom of the magnet frame is made slightly concave, and there is a rubber flange around the outer edge. Compressed air is piped into the hollow region at a gauge pressure of one or two atmospheres. The heavy magnet rises up a few millimeters and can then be delicately maneuvered.

The bottom of a 10-ton magnet might be 1 m $\times$ 2 m. The area is 2 $m^2$. If the gauge pressure (pressure above atmospheric) is 1 atm (half that in a car tire), the upward force is $2 \times 10^5$ N $= 20 \times 10^4$ N $=$ 20 metric tonnes. That's more than enough to float the magnet.

## WEIGHING YOUR CAR

You can determine the listed weight of your car by looking at the registration data. To check this out you can measure the footprints of your tires and their air pressure. After all, the car weight is balanced by the force of the compressed air in the tires.

You can measure the width of the "footprint" of a tire by placing a sheet of paper on the pavement and rolling the car forward until it rests on the paper. Then roll the car back and examine the print on the paper. The effective width of the tire will be less than the apparent width because there are gaps between the treads. Only the treads make contact with the pavement. You can measure the length of the footprint by sliding a sheet of paper or thin cardboard along the pavement and under the front of the tire until it can go no farther. Slide another sheet of paper from the back of the tire and then measure the distance between the two pieces of paper.

Here are data for a 1991 Lexus with new tires, parked on a slight grade to the left. Note that we are not using SI units. (It's hard to find a tire pressure gauge calibrated in N/$m^2$!) The area is in square inches,

**Table 1.1**
**1991 Lexus Automobile Tires**

| | *Width* | *Length* | *Area* | *Pressure* | *Support Force* |
|---|---|---|---|---|---|
| Left front | 4½ | 8¼ | 37.1 | 25 | 928 |
| Right front | 4½ | 7¾ | 34.9 | 25 | 872 |
| Left rear | 4½ | 5¾ | 25.9 | 25 | 648 |
| Right rear | 4½ | 6.0" | 27.0 | 25 | 675 |
| | | | | | 3120 |

the pressure is in pounds per square inches, and the force is measured in pounds.

Fortuitously, no doubt, the registration lists the weight of this car as 3133 lb. Note that the front of the car is heavier than the rear, and that the support provided by each tire depends on how level the road is.

## WELL AND WATER TOWER

How deep are wells? How high are water towers? In both cases, the answer depends on the pressure exerted by a column of water. Normal atmospheric pressure can be given in a variety of units: 14.7 lb/in$^2$, $1 \times 10^5$ N/m$^2$, 76 cm of mercury, or 10.3 m of water. These last two values specify the height of the liquid in a barometer tube that can be supported by atmospheric pressure. They also indicate the height to which a vacuum can pull a column of that liquid.

In a shallow well, a vacuum pump on the surface pulls up a column of water. As a practical matter, this system doesn't work well beyond about 25 ft. The pump has to be efficient for large volumes of water, with no air leaks. For deeper wells, a pressure pump is inserted at the bottom and shoves the water up.

Water towers, seen from afar, all appear to have the same height. The role of the high reservoir is to act as a constant pressure source, smoothing out the pulses of the pump that feeds it, or accommodating the variable demands of customers. The car battery serves the same function in a car, maintaining constant voltage. The pressure in a neighborhood distribution system must be able to provide water for at least a three-story house on the highest point in the region, requiring a pressure head of at least 10 m. Most towers are higher than that by a factor of 4 or 5, partly because pressure will drop along the line to remote customers. If the pressure head is much greater than 5 atm (75 lb/in$^2$), customers near the tower may find their pipe joints leaking.

## PRESSURE OF SUNLIGHT

We are being bombarded, day and night, by sunlight. It continually exerts a force of repulsion between Earth and Sun. Will it knock us off course?

The radiation power of the sun at the Earth's orbit is about 1.3 kW/m$^2$. If each photon carries 3 eV $= 5 \times 10^{-19}$ J, then the number of photons striking the Earth in one second in one square meter is

$$\frac{(1.3 \times 10^3 \text{ J/s} \cdot \text{m}^2)}{5 \times 10^{-19} \text{ J/photon}} = 3 \times 10^{21} \text{ photons/s} \cdot \text{m}^2.$$

The energy carried by a photon is $h\nu$ and the momentum it carries is $\frac{h\nu}{c}$. Since $F = \frac{\Delta \text{ momentum}}{\Delta \text{ time}}$ and $P = \frac{F}{A}$, the total pressure exerted by these photons is

$$P = \frac{(3 \times 10^{21})(5 \times 10^{-19})}{3 \times 10^8} = 5 \times 10^{-6} \text{ N/m}^2.$$

This may seem like a trivial amount, but the Earth's surface is large. Let's model the Earth as a flat disk with a radius of $6.4 \times 10^6$ m. Then the total force exerted on the Earth by sunlight is

$$F = (5 \times 10^{-6})\, \pi\, (6.4 \times 10^6)^2 = 6.4 \times 10^8 \text{ N}.$$

Since $10^4$ N, a metric ton, is about one American ton, the repulsive force of sunlight on the Earth is about 60,000 tons. Of course, that's very small compared with the gravitational attraction, which is

$$F_{\text{grav}} = G\frac{m_{\text{E}}M_{\text{S}}}{r_{\text{E-S}}^2} = (6.7 \times 10^{-11})\frac{(6 \times 10^{24})(2 \times 10^{30})}{(1.5 \times 10^{11})^2}$$
$$= 3.6 \times 10^{22}\ \text{N} = 3.6 \times 10^{18}\ \text{tons}.$$

Evidently, the force of sunlight does not present a threat.

## SQUASHED TENNIS BALL

When a golf ball or baseball or tennis ball gets hit, it flattens to an astonishing degree. Let's put in the numbers for a tennis ball. Drop it 1 m onto a hard surface, and it will bounce back about 63 cm. The speed with which it hits is

$$v_{\text{down}} = \sqrt{2gh} = \sqrt{2(9.8)(1)} = 4.4\ \text{m/s},$$

and the speed with which it starts to rise is $v_{\text{up}} = \sqrt{2(9.8)(0.63)} = 3.5\ \text{m/s}$. With a mass of 56 g, the change of momentum of the ball is

$$\Delta mv = 0.056\,[(-4.4\ \text{m/s}) - (+3.5\ \text{m/s})] = -0.44\ \text{kg}\cdot\text{m/s}.$$

The impulse on the ball is $\int F\,\text{d}t = \Delta(mv)$. If the collision were symmetric, with energy conserved, $F(t)$ would look like one of the lines in the diagram. The area under the curve is the impulse. To find the maximum force on the ball, we must calibrate the time axis. If the ball slows down with constant acceleration, its average speed in stopping would be 2.2 m/s. The distance it travels during that time depends on how much the ball is flattened. Let us assume that this distance is one-fourth of its diameter—about 1.5 cm. Then the time to flatten would be $(0.015\ \text{m})/(2.2\ \text{m/s}) = 6.8$ ms. Although we know that the approximation is crude because the collision is not symmetric, let us specify that the total collision time is 14 ms, and also assume that the curve is sinusoidal, as shown by the dotted line in figure 1.1.

$$F = F_{\text{max}} \sin \frac{2\pi}{T} t.$$

The area under the curve is $F_{\text{max}} \int_0^{T/2} \sin\frac{2\pi}{T}\,\mathrm{d}t = -F_{\text{max}}\frac{T}{2\pi}\cos\frac{2\pi}{T}t\,\Big|_0^{T/2}$ $= \frac{T}{\pi}F_{\text{max}}.$

$$F_{\text{max}} = \frac{\Delta(mv)}{T/\pi} = \frac{(0.44\ \text{kg}\cdot\text{m/s})}{(28\times10^{-3}\ \text{s})/\pi} = 49\ \text{N} = 11\ \text{lb}.$$

The weight of a standard tennis ball is only 0.56 N. The maximum squashing force is larger than this by a factor of 88, or 88 "g's."

For a different approximation, eschewing the calculus, find the area under the *triangular* curve.

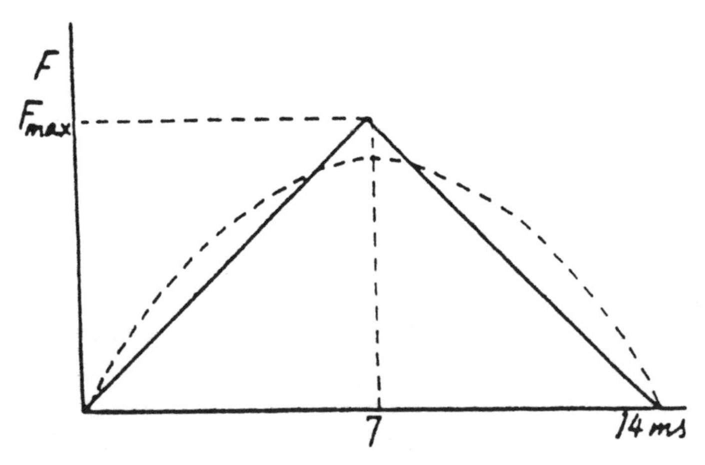


FIGURE 1.1 Approximation to force on a tennis ball as a function of time.

$$\Delta(mv) = (0.44\ \text{N} \cdot \text{s}) = {}^1\!/_2\, F_{\text{max}}(14 \times 10^{-3}\ \text{s})$$

$$\Rightarrow \quad F_{\text{max}} = \frac{0.44}{{}^1\!/_2(14 \times 10^{-3})} = 63\ \text{N}.$$

Note how $F_{max}$ depends on our assumption about the extent of squashing of the ball. If we use half the distance, then the collision time is cut in half, and $F_{max}$ would double.

## RECOIL

When you fire a gun, the bullet goes one way and the gun goes the opposite way. It's easy to see this effect when a person fires a pistol. The handle is below the barrel, and so the recoil produces a torque that makes the muzzle of the gun swing upward. It's also easy to feel the recoil if you have a powerful rifle against your shoulder.

Let's put numbers into the equation for momentum conservation, using data from a low-power "short" .22 gun. The slug has a mass of only 2 g; the muzzle velocity is 340 m/s. Compare this with the velocity of sound in air (343 m/s) or the speed of a passenger jet (270 m/s). The explosive powder also shoots out the muzzle in the form of a gas, but its mass is small for a gun of this size. The mass of a .22 rifle is about 3 kg. If the gun were floating, momentum conservation would require

$$(3\ \text{kg})V_{gun} = (0.002\ \text{kg})(340\ \text{m/s}) = 0.68\ \text{kg} \cdot \text{m/s}$$

$$V_{\text{gun}} = 0.23\ \text{m/s}.$$

The gun is not floating; it is braced against your arm pit. The recoil (the impulse) is given by $\int F dt$. Let's assume that the acceleration of the slug is constant, and the average speed while the slug is in the barrel is 170 m/s. The time it takes the slug to be shot out of a barrel that is ${}^2\!/_3$m long is

$$\frac{\frac{2}{3}\text{ m}}{170\text{ m/s}} = 3.9\text{ ms}$$

$$F_{\text{recoil}} = \frac{(0.68\text{ kg}\cdot\text{m/s})}{(3.9\times10^{-3}\text{ s})} = 170\text{ N}.$$

Compare this force with the weight of the gun, which is 30 N. For a few milliseconds, the recoil is exerting a force against your shoulder that is about six times the weight of the gun. With a "short" .22, you shouldn't feel any pain.

## FATAL IMPULSE

Before a car collision a driver has a momentum equal to $mv$; after the collision, the driver's momentum is zero. The vital question is, how long does it take to go from $mv$ to zero? Newton's second law can be written as

$$\int F\,\text{d}t = \Delta(mv) \qquad \left(\int F\,\text{d}t\text{ is called the impulse.}\right)$$

Suppose the car is traveling at 30 mph $\approx$ 14 m/s. The mass of the driver is 70 kg. Assume that the stopping distance is 1 m, allowing a generous amount of crushing of the front end and shoving the engine back. Then the average speed while stopping (assuming linear crushing) is 7 m/s. The total collision time is

$$\frac{1\text{ m}}{7\text{ m/s}} = 143\text{ ms}.$$

The diagram plots $F$ on the driver's body as a function of time for a couple of scenarios. The total areas under the curves must be the same and equal to

$$(70 \text{ kg})(14 \text{ m/s}) = 980 \text{ kg} \cdot \text{m/s}.$$

If the force is constant, then $\bar{F} = \frac{980 \text{ kg·m/s}}{0.143 \text{ s}} = 6.9 \times 10^3$ N.

Since the driver weighs only 700 N, the body must sustain an acceleration of almost 10 g's. Though it's for a short time, it's potentially lethal. There's no need to quibble, however, because our model assumes that the average force was applied in some average way to the whole body. Seat belts don't do that, but an air bag might save the driver.

The situation without air bag or seat belts is shown in the second curve of $F$ versus $t$ in figure 1.2. As the collision begins, the body flies forward, experiencing no force and no pain until it hits the steering

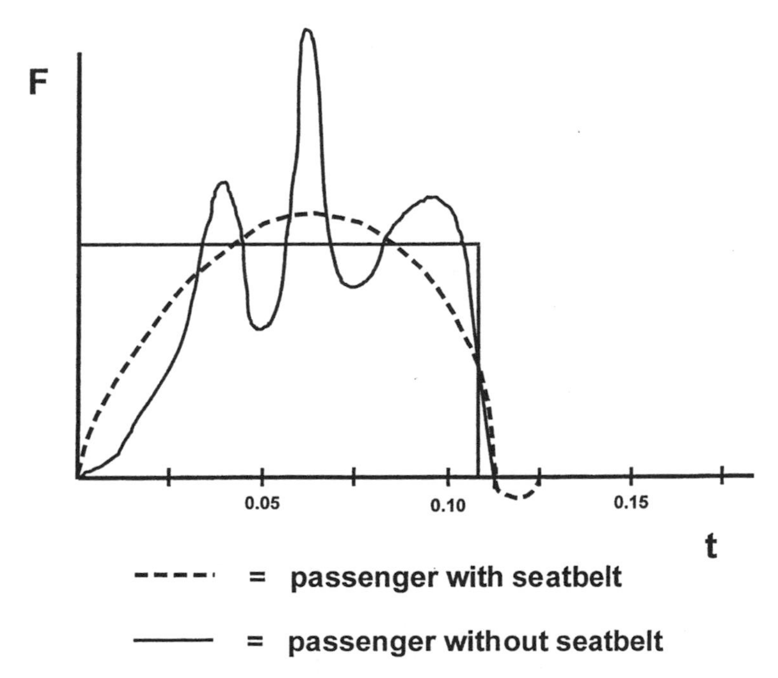


FIGURE 1.2 Force as a function of time for car and driver during collision with initial speed of 30 mph.

wheel. The force rises dramatically at that point but decreases as the wheel breaks or the body is pierced. The next rise in force is related to the unfortunate reaction of the head hitting the windshield. Regardless of how quickly the collision occurs, the areas under the curves must always equal $\Delta(mv)$. Common sense, civil law, and the laws of physics require you to wear a seat belt.

## RELATIVE GRAVITATIONAL WEAKNESS

All of us have experienced the feeling of heaviness. The gravitational force seems large because the mass of the Earth is so huge. It is of little comfort to realize that if we weigh 700 N in the Earth's gravitational field, the Earth weighs 700 N in our field. If two people, each of mass 70 kg, are 10 m apart, the attraction between them is

$$F = G\frac{m^2}{r^2} = 6.67 \times 10^{-11}\frac{(70\text{ kg})^2}{(10\text{ m})^2} = 3.3 \times 10^{-9}\text{ N}.$$

That's the weight of a mass of $3 \times 10^{-5}$ g in the Earth's field. (We assumed that Newton's equation was approximately valid even if the two people were not point objects or spheres.)

Although gravity dominates the celestial motions, it is infinitesimal in atomic particle interactions. Consider the relative strengths between the electric and gravitational attractions of an electron and proton in the Bohr atom.

$$\frac{F_{\text{electric}}}{F_{\text{grav}}} = \frac{kq_{\text{p}}q_{\text{e}}}{Gm_{\text{p}}m_{\text{e}}}$$

$$= \frac{(9 \times 10^9)(1.6 \times 10^{-19})^2}{(6.67 \times 10^{-11})(1.67 \times 10^{-27})(9.1 \times 10^{-31})} = 2.3 \times 10^{39}.$$

So, for most calculations in atomic or nucler physics, there's no need to worry about gravity.

# STRONG FIELDS

In the heart of an atom there are strong electric and magnetic fields. Consider, for example, the electric field strength at the classical Bohr orbit in a hydrogen atom.

$$E = k\frac{q}{r^2} = (9 \times 10^9)\frac{1.6 \times 10^{-19}}{(0.53 \times 10^{-10})^2} = 5.1 \times 10^{11}\ \text{V/m}.$$

The electric field at ground level from one of the high-voltage transmission lines is only about 1000 V/m.

The electric *potential* at that distance from the proton is less shocking. Considering the field strength, the potential is very small.

$$V = k\frac{q}{r} = (9 \times 10^9)\frac{(1.6 \times 10^{-19})}{(0.53 \times 10^{-10})} = 27\ \text{V}.$$

Notice that we have not assumed that the radius of the atom is the standard Å (Ångstrom) $= 1 \times 10^{-10}$ m. Instead, to find $r$, we solved the equation linking electrical attraction and centripetal force, applying the Bohr condition that requires an integral value of angular momentum.

$$k\frac{q^2}{r^2} = m\frac{v^2}{r} \quad \Rightarrow \quad r = k\frac{q^2}{mv^2} = kq^2\frac{mr^2}{(mvr)^2} = kq^2\frac{mr^2}{\hbar^2}$$

$$r = \frac{\hbar^2}{kq^2m} = \frac{(1.05 \times 10^{-34})^2}{(9 \times 10^9)(1.6 \times 10^{-19})^2(9 \times 10^{-31})}$$

$$= 0.53 \times 10^{-10}\ \text{m}.$$

Hydrogen is smaller than most atoms by a factor of 2. The ionization potential of hydrogen is only 13.6 eV. The electron is bound by the sum of the kinetic and potential energy: (+13.5 eV kinetic) − (27 eV potential).

The magnetic field at the center of the Bohr atom is produced by the electron orbiting the proton, producing a circular current.

$$i = \frac{ev}{2\pi r}.$$

The velocity is calculated from the same equation used to determine $r$.

$$v = \frac{\hbar}{mr} = \frac{(1.05 \times 10^{-34})}{(9 \times 10^{-31})(0.53 \times 10^{-10})} = 2.2 \times 10^{6} \text{ m/s}.$$

The magnetic field at the center of a circular loop is

$$B = \frac{\mu_o i}{2r} = \frac{\mu_o ev}{4\pi r^2} = 10^{-7}\frac{(1.6 \times 10^{-19})(2.2 \times 10^{6})}{(0.53 \times 10^{-10})^2} = 12.5 \text{ T}.$$

That's a strong magnetic field in the region of the proton. The proton must be affected by it.

Let's use a classical model to calculate the magnetic moment of the proton, which has an intrinsic angular momentum, $L$. This model is worked out in the item on "Magnet Strength" in chapter 6. For the proton,

$$\mathbf{m}_\text{p} = \frac{e}{2m}L = \frac{(1.6 \times 10^{-19})}{2(1.7 \times 10^{-27})}(1.05 \times 10^{-34})$$
$$= 5.0 \times 10^{-27} \text{ A} \cdot \text{m}^2 \text{ (classical model)}.$$

The magnetic moment of the electron would be larger by the ratio of proton mass to electron mass: 1836. Therefore, $\mathbf{m}_e = 9.3 \times 10^{-24}\text{A} \cdot \text{m}^2$ The measured magnetic moment of the electron agrees with this classical model, but the value for the proton is about three times greater than the one we calculated. Furthermore, the neutron, which has a net charge of 0, nevertheless has a large magnetic moment. Evidently the classical model breaks down for nucleons. The proton is not a simple point charge but has internal structure. The magnetic moment must be due to the circulation of the quarks and gluons that make up the proton. The experimental value for the magnetic moment of the proton is $\mathbf{m}_\mathrm{p} = 1.4 \times 10^{-26}\ \mathrm{A} \cdot \mathrm{m}^2$.

The calculation of the magnetic field produced by an orbiting electron is suspect. The electron has no rotational motion in the ground state. However, there is an alternative model in terms of the field produced by a magnetic dipole. The field along the axis at a distance $r$ from an electron magnetic dipole is

$$B = \frac{\mu_0}{2\pi}\frac{\mathbf{m}_e}{r^3} = (2 \times 10^{-7})\frac{(9.3 \times 10^{-24})}{(0.53 \times 10^{-10})^3} = 12\ \mathrm{T}.$$

That's the same field predicted by our first model. In the classical (Bohr) model, the proton can align itself with the magnetic field or be in the opposite direction. The resulting phenomenon agrees with these models. This phenomenon is described in "The 21 cm Line" in chapter 8.

## AIRPLANE LIFT

There is a perennial argument as to whether airplane lift is caused by the Bernoulli effect or by Newton's third law. One of the problems in appealing to Bernoulli is that the conditions required are not generally satisfied by the actual flying conditions. In an airplane, energy is fed into the system, whereas Bernoulli assumes energy conservation within the system. Furthermore, aside from waving hands, there is no elementary way to calculate the relative velocities, and thus pressures, of the air above and below the wing.

Newton's third law, however, must always be satisfied. For an airplane to stay up, a lot of air must be driven down. The down draft is obvious with a helicopter but must also exist with a fixed-wing plane. The weight of the plane is balanced by an upward force of the air on the plane; this upward force is the reaction to the downward thrust that the plane exerts on the air. (Fig. 1.3)

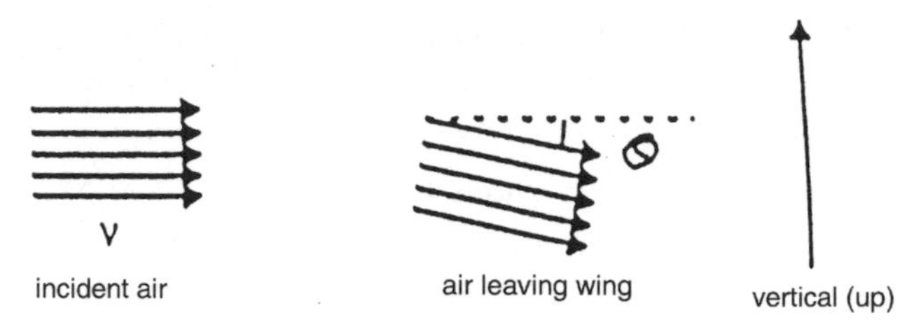


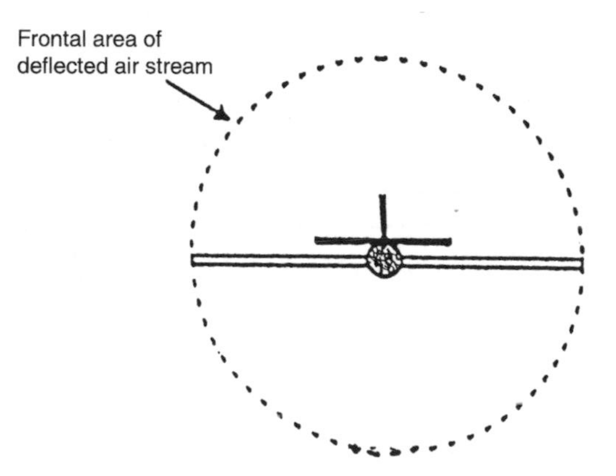


FIGURE 1.3 Air flow producing lift on airplane.

The thrust is $F = \frac{d(mv \sin\theta)}{dt} = v \sin\theta \frac{dm_{air}}{dt}$, where $v$ is the speed of the plane and $v \sin\theta$ is the speed downward of the deflected air. The air flows over the wing at constant $v$, and so $\frac{d(v \sin\theta)}{dt} = 0$.

Here are sample data for a light plane:

$$M = 1000 \text{ kg} \qquad v = 100 \text{ mph} = 45 \text{ m/s}$$

$$\rho_{air} = 1.3 \text{ kg/m}^3 \qquad \text{wing span} = 15 \text{ m}$$

The cross section of air affected is $\frac{\pi}{4}(15)^2$. That's the area of a circle with a diameter equal to the wingspan. The justification for this asumption is given in the reference at the end of this section. Note that a propeller would also disturb the air with a cross section equal to the circular area swept out. Consider, also, that the pressure pulse of a truck passing you on the highway extends ahead and to the sides of the truck.

For a trial calculation, let's assume that the angle of air deflection is equal to 10°. Sin 10° = 0.17.

The *volume* of air deflected downward per second is

$$\frac{\pi}{4}(15 \text{ m}^2)(45 \text{ m/s})(0.17) = 1.36 \times 10^3 \text{ m}^3/\text{s}.$$

The *mass* of the air deflected per second is $1.8 \times 10^3$ kg/s.

The resulting thrust is $v \sin\theta \frac{dm_{\text{air}}}{dt} = (45 \text{ m/s})(0.17)(1.8 \times 10^3 \text{ kg/s}) = 1.3 \times 10^4$ N.

This thrust is more than enough to support the plane's weight of 10,000 N. Note that this calculation is not at all based on a model of air molecules bouncing off the bottom of the wing. Indeed, most of the molecules never touch the wing at all. Their deflection is caused by a pressure pulse ahead of the advancing wing.

Scaling the argument for a B-747 works well. Here are the data:

$$M_{\text{laden347}} \approx 350 \text{ tons} \Rightarrow 3.5 \times 10^6 \text{ N} \qquad \text{Wingspan} = 64 \text{ m}$$

$$v = 300 \text{ m/s} \qquad \rho_{\text{air}} \text{ (at low altitudes)} = 1.3 \text{ kg/m}^3.$$

Since the lift is proportional to the square of the speed and to the square of the wingspan, the lift of the 747 will be greater than that of the small plane by a factor of $(\frac{300}{45})^2 \times (\frac{64}{15})^2 = 8.1 \times 10^2$. Therefore, the lift of the 747 would be $1.1 \times 10^7$ N ≈ 1100 tons. That is more than enough to keep the 747 aloft, even at an altitude of 8 km, where $\rho_{\text{air}} = 0.6$ kg/m$^3$ (and where the plane is lighter, having already used a lot of fuel).

The background of this calculation is given in "Flight without Bernoulli" by Chris Waltham in *The Physics Teacher* 36, (November 1998): 457.

Chapter

# 2

# Mechanics and Rotation

## TERMINAL VELOCITY

Galileo knew very well that all objects do not fall at the same rate in air. In his *Dialogue on Two New Sciences* Galileo described the experiment of dropping two balls of different weights from a high tower. (However, although he says that he has conducted the experiment, he does not place the event at the Leaning Tower of Pisa.) He points out that the heavier ball will be an "arm's span" ahead of the lighter ball but asks if, because of this slight discrepancy, the reader would choose the theory of Aristotle, who had the balls separated by half the height.

Air friction for most human-size objects is proportional to the square of the velocity. The equation of motion including air friction is

$$mg - kv^2 = ma.$$

If the drag coefficient, $k$, equals 0, then $g = a$, and we have free fall. If $k$ does not equal 0, then the friction term rapidly grows, decreasing the acceleration to zero. At the terminal velocity for falling, the weight is just equal to the air friction force.

$$\text{When } a = 0, \quad mg = kv^2 \quad \text{and } v_{\text{terminal}} = \sqrt{\frac{mg}{k}}.$$

Note that terminal velocity does depend on the mass of the falling object and that the smaller the streamline constant, $k$, the larger the terminal velocity. The constant, $k$, depends on a streamline coefficient, $C$, the density of air, $\rho$, and the surface area, $A$, of the object.

$$k = {}^1\!/\!{}_2\, C\, A\, \rho.$$

The value of $C$ for a smooth sphere is 0.5. For a standard baseball with a mass of 142 g and a diameter of 7.3 cm,

$$v_{\text{T}} = \sqrt{\frac{2(0.142 \text{ kg})(9.8 \text{ N/kg})}{0.5(42 \times 10^{-4} \text{ m}^2)(1.3 \text{ kg/m}^3)}} = 32 \text{ m/s} = 71 \text{ mph}.$$

The actual terminal velocity of a baseball is about 95 mph. Baseballs are not smooth. The roughness enhances a turbulent flow of air, which twirls into a smaller wake (and creates less resistance) than part laminar, part turbulent flow.

We can rewrite the equation for $v_{\text{T}}$ of smooth spheres in terms of the density, $D$, of the sphere. The mass of the sphere becomes $\frac{4}{3}\pi r^3 D$.

$$v_{\text{T}} = \sqrt{\left(\frac{8g}{3C\rho}\right) r D} = 6.3\sqrt{rD}.$$

The density of iron is $7.9 \times 10^3$ kg/m$^3$. The density of a standard baseball is $7.0 \times 10^2$ kg/m$^3$ (note that a baseball can float in water). A sphere of iron with the same radius as a baseball would have a terminal velocity of $(32 \text{ m/s})\sqrt{\frac{(7.9\times10^3)}{7.0\times10^2}} = 108 \text{ m/s}$.

A golf ball has a radius of 2.1 cm and a density of $1.2 \times 10^3$ kg/m$^3$. (As every golfer knows, golf balls sink!) Our formula gives a terminal velocity of 32 m/s, the same as for a baseball. The actual terminal velocity is 40 m/s, a little less than that of a baseball. Once again, the golf ball is not smooth, reducing its air friction at high velocities.

A ping pong ball is smooth and has a very small density. The radius is 1.9 cm and the density is 94 kg/m$^3$. Our formula yields 8.4 m/s = 19 mph, in remarkable agreement with experiment.

The terminal velocity of a human is from 100 to 200 mph depending on the size of the human. The standard terminal speed for a *successful* parachutist is 15 mph. If landing speed is much greater, he breaks his ankles; much less, and he drifts into trees. Heavier people need larger parachutes.

Another example of the effect of air friction is commonly observed with cars. Depending on the type of car, the major friction at speeds up to about 30 mph is caused by tires flexing against the road. As the speed increases, the $v^2$ term due to air friction rapidly dominates. Finally, the car reaches its terminal speed no matter how powerful the engine.

To determine if you must take air friction into account, consider as a rule of thumb that in a fall through air starting at rest, an object's speed remains proportional to time within 10% for speeds up to 80% of terminal velocity.

## THE UNLIKELY GAME OF BASEBALL

The duel between pitcher and batter is settled in a fraction of a second, and within a 3° precision in delivery of the pitch. The distance between pitcher and batter is 60 $^1/_2$ ft, in American units—and, after all, it is an American game. Nevertheless, for ease of calculation we shall use SI units, which makes the distance between pitcher and batter 18.44 m. For a speed of 90 mph (40.23 m/s), the time between delivery and crossing the plate is

$$\frac{18.44 \text{ m}}{40.23 \text{ m/s}} = 0.458 \text{ s}.$$

The batter's reflex time is at least $^1/_5$ s, leaving him about $^1/_4$ s to decide whether and how to swing.

If the pitcher throws precisely horizontally, the ball will drop to about the bottom of the strike zone, depending on the height of the batter and the judgment of the umpire.

$$\Delta y = -\frac{1}{2} g \, \Delta t^2 = -4.9(0.458)^2 = 1.03 \text{ m}.$$

The vertical distance between the ball leaving the pitcher's hand and the knees of the batter is about a meter. (In the next approximation we should take into account the height and style of the pitcher, and the height of the pitcher's mound.)

If the pitcher wants to reach the top of the strike zone, then he must throw at an angle slightly above the horizontal so that the ball returns to the same height from which it started.

$$(v_0 \sin\theta)\Delta t = \frac{1}{2} g(\Delta t)^2 \quad \Rightarrow \quad 40.23 \sin\theta = 4.9 \times 0.458$$
$$\sin\theta = 0.0558 \quad \Rightarrow \quad \theta = 3.2°.$$

Of course, there is no need for baseball players to do these calculations. Real baseballs move through air, suffering drag and spin interactions. The terminal velocity of a baseball is about 95 mph (see the previous section, "Terminal Velocity"). Since throwing and batting speeds are this great or greater, air friction must play a major role. But these first approximations make it seem astonishing that the game can be played at all.

## HUMAN LEVERS

Levers involve a trade-off between force and the distance through which the force is exerted. The mechanical advantage of a lever is the ratio of output force to input force. Usually we use a lever to increase the output force at the expense of exerting the input force through a longer distance. Interestingly, the human body has a number of levers that work in just the opposite way. They have a fractional mechanical advantage.

Figure 2.1 shows the geometry of bones and muscles in the arm. Muscles can only contract and pull; they cannot expand and force outward. Therefore, we have biceps to pull the hand up and triceps to pull the hand back down. The fulcrum around which the lower arm turns is located about 5 cm from where the biceps are attached. The hand has a lever arm of about 35 cm from the fulcrum. This geometry produces a mechanical advantage of 1/7. If the hand holds a stone weighing 10 N, the biceps must exert a force of 70 N.

Why should the human body have developed such an inversely simple machine? Notice that as the biceps contract 5 cm, the hand with the 10 N stone rises 35 cm. If the contraction takes 0.1 s, the stone can leave the hand at 3.5 m/s. The human body trades strength for speed.

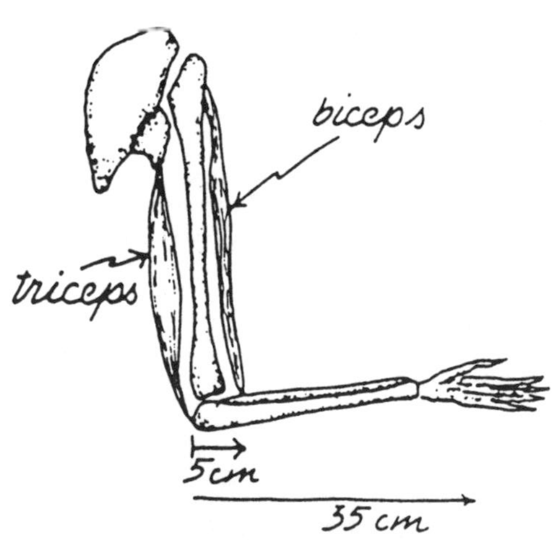


FIGURE 2.1 Leverage of human arm.

# ANGULAR MOMENTA

Objects appear to be spinning or are in orbit at every range of magnitude, from the subatomic realm to the domain of the galaxies. A rotating object's angular momentum is $I\omega$, where $I$ is the moment of inertia about the axis and $\omega$ is the rotational velocity in radians/second. Let's calculate the angular momentum of objects in three different ranges of size. We should compare these angular momenta in terms of some common unit. Although, as we shall see, nature has a basic unit of angular momentum, no special name is given to the SI, or human-made unit. In terms of kilograms, meters, and seconds, the unit of angular momentum is 1 kg $m^2$/s.

First, let's start with an object that is human sized. An ordinary bicycle wheel has a mass of about 2 kg and a radius of 30 cm. If we make the approximation that all of the mass is at the rim, then a wheel traveling at 10 mph ($4\,{}^1\!/_2$ m/s) has a spin (an intrinisic angular momentum about its axis) of

$$mvr = (2\text{ kg})(4.5\text{ m/s})(0.30\text{ m}) = 2.7\text{ kg m}^2/\text{s}$$

If the wheel is rolling along the road, there is an equal but separate value of $mvr$ with respect to an instantaneous axis where the rim touches the road.

Second, in the Bohr model of the atom, proposed by Bohr in 1913 and still very useful, the valence electron in an atom travels in a circular orbit around the central nucleus. The mass of an electron is about $1 \times 10^{-30}$ kg; the radius of an atom (and therefore the radius of the orbit of the outer electron) is about $1 \times 10^{-10}$ m, and the speed of the electron is about $1 \times 10^{6}$ m/s. (These values are calculated in the section "Valence Electrons" in chapter 10) . *On the basis of this model*, the angular momentum of such an electron is

$$\begin{aligned} mvr &= (1 \times 10^{-30}\text{ kg})(1 \times 10^{6}\text{ m/s})(1 \times 10^{-10}\text{ m}) \\ &= 1 \times 10^{-34}\text{ kg m}^2/\text{s}. \end{aligned}$$

That may seem like an unimaginably small quantity. Compare it with the angular momentum of a bicycle wheel. But look at the number—we have obtained the (reduced) Planck's constant, $h/2\pi = \hbar$,which is the quantum unit of angular momentum!

Finally, let us calculate the angular momentum of something big. The Earth spins on its axis with a frequency of one revolution per day. If the Earth's density were uniform, which it is not, the moment of inertia would be

$$I = \frac{2}{5} MR^2 = 0.4\, MR^2.$$

Because the core of the Earth is more dense than the mantle, the actual moment of inertia is 0.3444 $MR^2$. The angular momentum is

$$I\omega = (0.344)(6 \times 10^{24}\ \text{kg})(6.4 \times 10^6\ \text{m})^2 \frac{2\pi\ \text{rad}}{24 \times 3600}$$
$$= 6.1 \times 10^{33}\ \text{kg m}^2/\text{s}.$$

In powers of 10, this quantity of angular momentum is about as large in powers of 10 as the size of the angular momentum of the valence electron is in the other direction. Actually, the angular momentum of the spinning Earth is small compared with other angular momenta in the solar system.

## THE TWIRLING SKATER

One of the most dramatic demonstrations of the conservation of angular momentum is the twirling skater. With no apparent effort a dancer or skater can go from standing still into dizzying rotation. Careful observation shows that the spin starts slowly with the skater's arms outstretched, then the arms are drawn in and the rapid spinning begins. To see if angular momentum is conserved, put in the numbers.

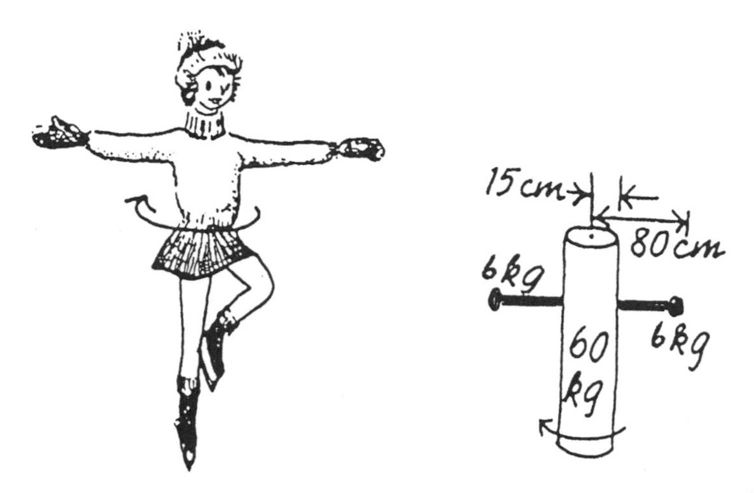


FIGURE 2.2 Approximation model of a twirling skater.

The moment of inertia of a complex system is $I = \sum mr^2$. We model the skater's body by a cylinder of radius 15 cm and a mass of 60 kg. As shown in figure 2.2, each outstretched arm of the model has a mass of 6 kg and a radius of 80 cm. The moment of inertia of a solid cylinder (the body) is $\frac{1}{2}mr^2$. The moment of inertia of a rod of length, $L$, rotating about one end (an arm), is $\frac{1}{3}mL^2$. The crude model starts out rotating with a frequency of one revolution every 2 seconds; $\omega = 3.1$ rad/s. The total angular momentum (including two arms) is

$$[{}^1\!/_2(60\text{ kg})(0.15\text{ m})^2]\omega + [({}^{12}\!/_3\text{ kg})(0.80\text{ m})^2]\omega$$
$$= [0.68]\omega + [2.6]\omega = 10\text{ kg}\cdot\text{m}^2/\text{s}.$$

Note that the angular momentum of the much lighter arms and hands is about four times the angular momentum of the core body.

Next, the skater brings her arms and hands down beside her body. The angular momentum stays the same, but the radius of the hands and arms is greatly reduced.

$$10 \text{ kg} \cdot \text{m}^2/\text{s} = [0.68]\omega + [(12 \text{ kg})(0.15 \text{ m}^2)]\omega$$

$$= 0.68\ \omega + 0.27\ \omega = 0.95\ \omega$$

$$\omega = 11 \text{ rad/s}.$$

The angular frequency has increased by over a factor of 3. The twirler is now rotating at about two revolutions per second.

## ROTATING WATER PAIL

A standard demonstration in an introductory physics course is to fill a pail of water and then swing it in a vertical plane. If the period of the rotation is sufficiently short, no one gets wet, though students worry.

First, let's calculate the limiting period and then raise a possible student objection. Transferring to the rotating reference frame, in order that the water not spill, the centrifugal force on a chunk of water at the top of the swing must be equal to or greater than the weight of that water.

$$m\frac{v^2}{r} = m\ \omega^2\ r = m\frac{4\pi^2}{T^2}r = mg$$

$$T = 2\pi\sqrt{\frac{r}{g}}.$$

Note that this period is the same as the period of a simple pendulum that has length $r$. For a swinger with a normal arm length and using a standard pail, $r = 1.2$ m (from shoulder down to the water surface).

$$T = 2\pi\sqrt{\frac{1.2}{9.8}} = 2.2 \text{ s}.$$

Therefore, the frequency is 0.45 Hz.

The period is long enough so that the action is almost slow motion. However, some bright student might object that the pail is upside down for so short a time that the water does not have time to fall out. Consider a chunk of water on the surface. For one-sixth of the period, it is within 30° of the top of the swing, and the downward component of the gravitational force is within 13% of $g$. The distance the chunk of water can fall is

$$y = \frac{1}{2}at^2 = {}^1\!/_2 9(\frac{1}{3}\ \mathrm{s})^2 = \frac{1}{2}\ \mathrm{m}.$$

If the water fell down that much it would be a very wet demonstration. Evidently, the explanation of the effect depends on dynamics, not kinematics.

## GRANDFATHER CLOCK

According to legend, Galileo measured the period of a swinging church chandelier by timing it with his pulse. He thought that the period was independent of the amplitude, which is a reasonable observation considering the circumstances. We now know that this is approximately true only for small angles. The differential equation for a simple pendulum is

$$\frac{\mathrm{d}^2\theta}{\mathrm{d}t^2} = -\frac{g}{L}\sin\theta \approx -\frac{g}{L}\theta \text{ (for small } \theta).$$

For small $\theta$, $T = 2\pi\sqrt{\frac{L}{g}}$, independent of angle and the mass of the bob. For $L = 1$ m, $T = 2$ s. For $T = 1$ s, $L = 0.25$ m (about 1 ft). For $L = 100$ m, $T = 20$ s.

A grandfather clock does not use a simple pendulum. The pendulum arm consists of several rods, usually of two different metals, arranged to maintain constant length in spite of temperature changes. The differential equation for a physical pendulum is

$$\frac{\mathrm{d}^2\theta}{\mathrm{d}t^2} = -\frac{\text{torque}}{\text{moment of inertia}} = -\frac{mg\,\rho\sin\theta}{I}.$$

The radial distance from the axis to the center of mass is $\rho$. For a grandfather clock, $\theta$ is always small, so

$$\frac{\mathrm{d}^2\theta}{\mathrm{d}t^2} = -\frac{m\,g\,\rho}{I}\theta.$$

The period is $T = 2\pi\sqrt{\frac{I}{m\,g\,\rho}}$. For a meter stick, swinging from one end, $I = \frac{1}{3}mL^2$, and

$$T = 2\pi\sqrt{\frac{\frac{1}{3}mL^2}{mg\frac{1}{2}L}} = 2\pi\sqrt{\frac{\frac{2}{3}L}{g}} = 1.6\text{s}.$$

How small is a "small" angle? The complete expression for the period of a simple pendulum is

$$T = 2\pi\sqrt{\frac{L}{g}}\left(1 + \frac{1}{4}\sin^2\left(\frac{\theta_{\mathrm{M}}}{2}\right) + \frac{9}{64}\sin^4\left(\frac{\theta_{\mathrm{M}}}{2}\right) + \cdots\right).$$

For $\theta_{\mathrm{M}} = 30°$, the first correction term adds 0.017, or 1.7%. The period of a simple pendulum is independent of amplitude within 1% for angles up to 23°.

## ARM AND LEG PENDULUMS

When we walk, our arms swing back and forth, in resonance with our legs. Which extremities determine the frequency of swinging? Let's calculate the natural frequencies of arms and legs and see if the results agree with experiment. We will model the arms as being solid cylinders of constant radius. The moment of inertia of a uniform cylinder swinging about one end is $I = \frac{1}{3}ML^2$. The period of a physical pendulum is $T = 2\pi\sqrt{\frac{I}{Mgd}}$. In this formula, $M$ is the mass and $d$ is the distance from fulcrum to center of mass. For a uniform cylinder,

$$T = 2\pi \sqrt{\frac{\frac{1}{3}ML^2}{Mg\left(\frac{1}{2}L\right)}} = 2\pi\sqrt{\frac{\frac{2}{3}L}{g}}.$$

For an arm of length 66 cm, $T = 1.3$ s. Try it and see! Let your arm swing freely and let someone else time ten swings.

Since the legs are longer than the arms, the natural period should be longer. Length, for instance, might be 86 cm, yielding $T = 1.5$ s. To test this experimentally, stand with one foot on a stair step and let the other leg swing freely. There's a catch, however. While the arm can reasonably be modeled as a uniform cylinder, the leg's distribution of mass is not at all uniform over the length. To some extent, the foot at a large radius compensates for the greater thickness of the upper thigh. For most people, the simple model and calculation yield a natural period that agrees roughly with measurement. Furthermore, this natural frequency of oscillation is the frequency of motion in casual walking. It takes more energy to walk faster or slower. That's why it is tiring for a tall person to walk hand in hand with a short person—the natural period is proportional to the square root of the leg length.

The arm and leg oscillations are coupled. For many people, the arm oscillation drives the leg movement during normal walking. You can even observe a quarter phase difference between arm and leg oscillation, which is to be expected when a system is being driven at resonance. It is also why carrying a weight in each hand is tiring—it increases the moment of inertia of the arms and so increases their natural period, which forces an increase in the period of leg oscillation, which tires the legs.

Discussions of these phenomena can be found in *The Physics Teacher*: **14** (September 1976): 360 (by R. Prigo); **14** (September 1976): 360–61 (by C. H. Bachman); and **35** (September 1997): 372–76 (by A. Domont and C. Waltham).

# PRECESSION OF A BICYCLE WHEEL

A bicycle wheel with handles makes a great demonstration device for the phenomena of angular momentum and gyroscopic motion. The ball bearings produce very little friction. To demonstrate precession, hang the wheel from a cord tied to the handle, as shown in figure 2.3. If the wheel is not spinning, it will simply flop over so that the plane of the wheel is almost horizontal. Now start the wheel spinning with its plane vertical. It wlll precess around the axis of the cord.

Here are some reasonable data. Assume all of the mass (2 kg) is concentrated at the outer radius, $r = 30$ cm. Tie the cord on the

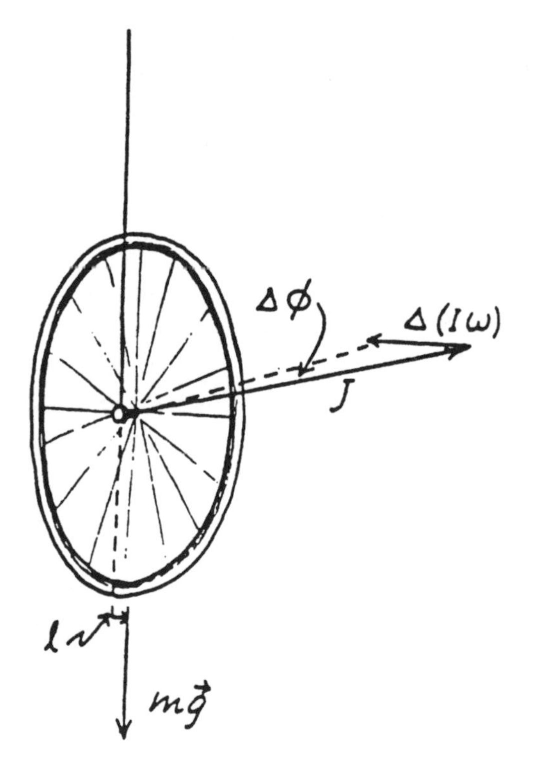


FIGURE 2.3 Geometry of precessing bicycle wheel suspended by a cord at a distance of $\ell$ from the center of gravity.

handle, 6 cm from the center of mass. Spin the wheel at $\omega = 10$ rad/s. The period is $T = 0.6$ s. Call the (spin) angular momentum, $J$.

$$J = I\,\omega = mr^2\omega = (2\text{ kg})(0.3\text{ m})^2(10\text{ rad/s}) = 1.8\text{ kg}\cdot\text{m}^2/\text{s}.$$

The frequency of precession is

$$\omega' = \frac{\mathrm{d}\phi}{\mathrm{dt}} = \frac{\text{torque}}{\text{angular momentum}} = \frac{mg\,\ell}{J}$$

$$= \frac{(2\text{ kg})(9.8\text{ N/kg})(0.06\text{ m})}{1.8\text{ kg}\cdot\text{m}^2/\text{s}} = 0.65\text{ rad/s}.$$

The period of the precession is $T = \frac{2\pi}{\omega'} = 9.7$ s. Both the rotational period and the precession period are easy to measure to within 10%.

## CAR SPRINGS

Car springs should provide a ride that is soft enough to be comfortable and yet not let the car bounce up and down too much. Springs and shock absorbers are complicated devices on modern cars, and each wheel is separately suspended. However, for a first-approximation model, let us assume that a 2000 kg car (about 2 tons) is supported by a single spring (perhaps the result of four in parallel). The spring might compress 20 cm under a full load. (Consider how high the car rises before the wheels clear the road when you are jacking the car to change a tire.) The spring constant would then be

$$k = \frac{F}{y} = \frac{(2\times 10^4\text{ N})}{0.2\text{ m}} = 1\times 10^5\text{ N/m}.$$

If the car hits a bump, it will vibrate with a period

$$T = 2\pi\sqrt{\frac{m}{k}} = 2\pi\sqrt{\frac{(2\times 10^3\text{ kg})}{(1\times 10^5\text{ N/m})}} = 0.9\text{ s}.$$

Within a factor of 2, that time agrees with the actual vibration period of most cars. If the shocks are working properly, the vibration will be damped within one or two periods.

## CAROUSELS AND ROTORS

Amusement park visits are popular with high school classes. The students measure various quantities of the rides and check the quantitative analysis with the effects they feel. Here are some sample data for a carousel—a merry-go-round of 7 m radius and a 25 s period of rotation.

The centrifugal field experienced by a rider at the outer radius is

$$\frac{v^2}{r} = \omega^2 r = \frac{4\pi^2}{T^2} r = \frac{40}{625} 7 = 0.45 \text{ N/kg}.$$

Compare this field strength with the gravitational field, $g = 9.8$ N/kg. A plumb bob would hang at an angle from the vertical whose tangent is $(0.45)/(9.8) = 0.046$. The angle is 2.6°. That's a barely noticeable effect. However, if you walk around while on the merry-go-round there's another force to contend with. If you are walking at 2 m/s (about 4 mph), the Coriolis field strength is

$$2\, \vec{v} \times \vec{\omega} = 2\ (2 \text{ m/s}) \left(\frac{2\pi}{25}\right) = 1 \text{ N/kg}.$$

The force would be perpendicular to your velocity and to the axis of rotation. If you walk radially, the force will be tangential, and that field strength is large enough compared with $g$ to make walking difficult, or at least surprising.

There's another device, called a rotor, that is common in most amusement parks. The radius is smaller than that of a merry-go-round and the rotation speed is much greater. Typical values are a 2 m radius and a 1.7 s period of rotation.

The riders stand, lying slightly back against the inside wall. When the rotor gets up to full speed, the bottom floor of the device retracts, leaving the riders plastered against the inner wall, held up only by the centrifugal force and the friction between them and the wall.

In this case, the centrifugal field strength is

$$\frac{4\pi^2}{T^2}r = \frac{40}{2.9}2 = 27.6 = \text{almost}\ \ 3\ \text{g's}.$$

The force exerted on the body by a field of 3 g's is borderline painful, so the ride does not last long. The tangent of the angle from the vertical is $\frac{3\ \text{g}}{1\ \text{g}} = 3$. A plumb bob would hang at an angle of 70° from the vertical. Riding on a rotor is clearly an exercise for young people, and not for adults!

## BANKED ROAD

When a curved road is properly constructed, it is banked to reduce slippage up or down. The geometry and variables are shown in figure 2.4. We can analyze the dynamics in the laboratory (road) frame or in the frame of the circling car. Figure 2.4a shows the first case, 2.4b the second. Of course, the two systems agree about the proper angle of banking. In the frame of the circling car, the vector sum of the normal force from the road, the weight, and the centrifugal force add up to zero. In that frame, the car is not accelerating. In the road reference frame, the components of the normal force provide the force to balance the weight and the necessary centripetal force to make the car follow a circular path. In either case:

$$N\sin\theta = \frac{mv^2}{r}; \quad N\cos\theta = mg; \quad \tan\theta = \frac{v^2}{rg}.$$

For a cloverleaf turn with radius of $\frac{1}{5}$ mile = 320 m, and for a design speed of 30 mph = 14 m/s,

$$\tan\theta = \frac{(14\ \text{m/s})^2}{(320\ \text{m})(9.8\ \text{m/s}^2)} = 0.063 \quad \Rightarrow \quad \theta = 3.6°.$$

a)

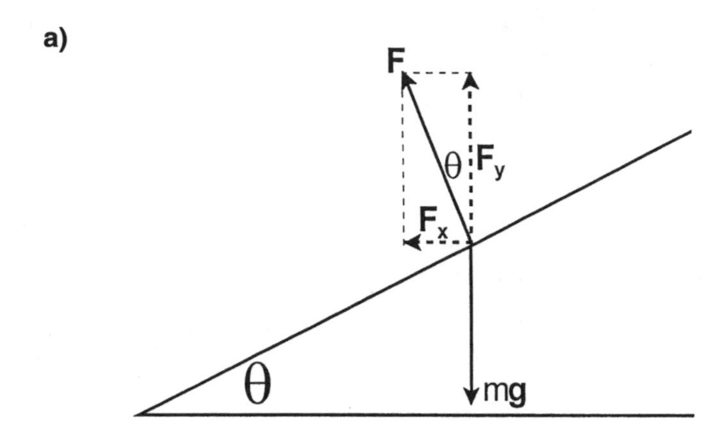


$$F_y = F\cos\theta = mg$$

(the vertical component of the force exerted by the road is equal to the weight of the car)

$$F_x = F\sin\theta = \frac{mv^2}{r}$$

(the horizontal component exerted by the road provides the centripetal force)

$$\therefore \tan\theta = \frac{v^2}{rg}$$

b)

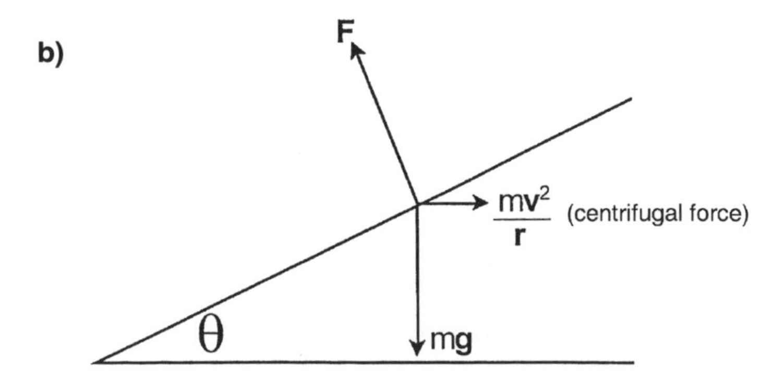


FIGURE 2.4 Geometry of forces acting on a car on a banked road.

This may seem like a small angle, but on a road it's a considerable slope. Consider that an incline of 10° is a very steep hill. If the speed were to double, the tangent (and hence, in the linear range) the angle would increase by 4, bringing the angle up to 14°. As a practical matter, no road would be banked by more than a few degrees.

## CONICAL PENDULUM

Using a string and bob, you can make the standard pendulum travel in a conical path. Figure 2.5 shows the geometry and labels the variables. In determining the direction of the radial force, we can choose to be either in the rotating reference frame or in the laboratory frame. The figure is labeled for the rotating frame, with a centrifugal force pointing outward radially. In this case, the three forces acting on the bob must cancel, since in the rotating reference frame the bob is not accelerating. In the laboratory frame, the unbalanced centripetal force

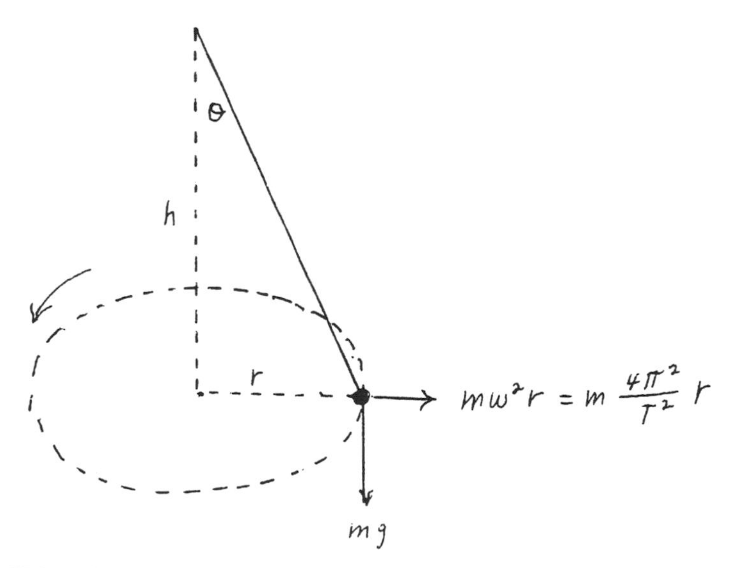


FIGURE 2.5 Geometry of forces acting on a conical pendulum.

would be necessary to continually force the bob to follow a circular path.

To produce good experimental results, mark a circle on the floor or table, and then hold the upper end of the string directly above the center of the circle and twirl the bob in a horizontal plane. It is surprisingly easy to produce a circular orbit that matches the one drawn on the surface. Keep the moving bob only a constant short distance above the drawn circle, thus determining the height, $h$. Have a partner measure the time for ten periods.

$$\tan\theta = \frac{r}{h} = \frac{(4\pi^2/T^2)r}{g} \qquad \Rightarrow \qquad T = 2\pi\sqrt{\frac{h}{g}}\,.$$

Notice the similarity between the formula for a conical pendulum and that for a simple pendulum:

$$T = 2\pi\sqrt{\frac{L}{g}}\,.$$

Notice also that the period of the conical pendulum is independent of $\theta$, $r$, and $L$, as long as $h$ remains constant.

Suppose $h = 10$ cm, $T = 2\pi\sqrt{0.01} \approx 0.6$ s. If $h = 100$ cm, $T = 2\pi\sqrt{0.1} \approx 2$ s. If $h = 50$ cm, $r$ can have any value, but if $r = 10$ cm, $\theta = 45°$ (an easy orbit to maintain). If $h = 10$ cm and $r = 100$ cm, $\theta = 84°$ (a very difficult orbit to maintain). If $h = 100$ cm and $r = 10$ cm, $\theta = 6°$.

Chapter

# 3

# Sound and Waves

## SPEED OF SOUND

Sound is carried by the motion of atoms or molecules, either dashing about in a gas or vibrating in a solid or liquid. In a gas, a sudden increase in density (a pressure pulse) spreads out as the molecules travel and collide with one another. This sound pulse must travel more slowly than the average speed of the molecules. That average (root mean square) molecular speed is derived from the expression for molecular kinetic energy:

$$\frac{1}{2}mv^2 = \frac{3}{2}kT \qquad \Rightarrow \qquad v = \sqrt{\frac{3kT}{m}}.$$

The Boltzmann constant is $k = 1.38 \times 10^{-23}$ J/molecule · K. The mass of a nitrogen molecule is $4.7 \times 10^{-26}$ kg. At room temperature of 293 K,

$$v = 5.1 \times 10^2 \text{ m/s}.$$

The speed of sound in air at room temperature is about 340 m/s, which is about two-thirds the average speed of the nitrogen molecules. If the speed of sound is always two-thirds of the mean molecular speed, then the speed of sound in helium should be

$$\frac{2}{3}(5.1 \times 10^2)\sqrt{\frac{28}{4}} = 900 \text{ m/s}.$$

The measured speed of sound in helium is 980 m/s. Evidently there are some extra considerations that we have not yet taken into account.

Newton, also, had trouble computing the speed of sound in air. Instead of appealing to molecular motion, which people didn't know about in those days, Newton used the general formula for the speed of a wave. It is equal to the square root of a restoring force term divided by the density of the medium. In the case of air, this relationship is

$$v = \sqrt{\frac{B}{\rho}}, \text{ where } B \text{ is the bulk modulus.}$$

The value of the bulk modulus of air depends on whether the compression is isothermal or whether it takes place so rapidly than heat does not escape the compressed region. For isothermal changes, $B = P$, the pressure. For the rapid adiabatic changes, without heat exchange from one region of the wave to the next, $B = \gamma P$, where $\gamma$ is the ratio of specific heats. Newton guessed wrong and used the isothermal value. For a gas such as air, composed of diatomic molecules, $\gamma = \frac{7}{5}$, and

$$v = \sqrt{\frac{\frac{7}{5}(1 \times 10^5 \text{ N/m}^2)}{1.29 \text{ kg/m}^3}} = 330 \text{ m/s},$$

which is correct at STP.

Now we see what went wrong with our simple approximation to find the speed of sound in helium. The ratio of specific heats, $\gamma$, for monatomic gases is $\frac{5}{3}$. We must multiply our previous result by an extra factor of $\sqrt{\frac{5/3}{7/5}} = 1.09$. Sure enough, $900 \times 1.09 = 980$, which is the measured speed of sound in helium.

In a solid, the compressions and rarefactions of sound are transmitted by the motion of atoms vibrating in their local potential wells. The response time for the increase of pressure must be about half the period of oscillation, typically about $10^{-13}$ s. In that time, the signal must travel a distance equal to the radius of the atom. The speed then is

$$v = \frac{\text{atomic radius}}{\frac{1}{2}/(\text{vibration frequency})}$$
$$= 2(1 \times 10^{-10}\ \text{m})(1 \times 10^{13}\ \text{Hz}) = 2 \times 10^{3}\ \text{m/s}.$$

This is the order of magnitude of the speed of sound in solids. The vibration frequency is inversely proportional to the square root of the mass of the atom, and so the speed of sound should be greater for solids low in the periodic table. That's generally true, although other factors enter, such as the strength of the atomic "spring constant." Diamond, for instance, has a large spring constant and a low atomic mass, and so the speed of sound in diamond is very large. In the case of lead, however, conditions are just the opposite and the speed of sound is relatively low. There are other complications as well. For instance, the sound disturbance can be compressional or transverse, with different speeds for each.

## TSUNAMIS AND RIPPLES

The velocity of a wave is usually given in terms of the square root of the ratio of a restoring force term to an inertial term. For instance, the velocity of a pulse in a stretched rope with tension $S$ and linear density $\mu$ is

$$v = \sqrt{\frac{S}{\mu}}.$$

If the tension in the rope is 100 N, and the linear density is 0.25 kg/m, then the velocity of a pulse in the rope is 20 m/s.

This formula is derived under several constraints. It is assumed that the speed is independent of the shape or amplitude or frequency of the pulse, and that the response of the rope is linear—that is, twice the force produces twice the displacement.

The motion of breaking water waves is much too complicated for easy analysis. The motion of water *ripples*, however, can be described with a single formula. Here is the formula, as derived in a paper by Kuwabara, Hasegawa, and Kono ("Water Waves in a Ripple Tank," *Am. J. Phys.* 54 [1986]: 1002):

$$v = \sqrt{\left(g\,\bar{\lambda} + \frac{\gamma}{\rho\,\bar{\lambda}}\right)\tanh\frac{D}{\bar{\lambda}}}\,.$$

The water rises as the wave passes by, and thus the velocity depends on the gravitational field strength, $g$; the density of the water, $\rho$; the surface tension, $\gamma$; the depth of the water, $D$; and the reduced wavelength, $\bar{\lambda} = \lambda/2\pi$.

In limiting cases, the formula is simpler than it looks. The hyperbolic tangent has the convenient property of being approximately 1 for $D/\bar{\lambda} > 1$, and being equal to $D/\bar{\lambda}$ for $D/\bar{\lambda} < 1$. Let's take the case for the reduced wavelength short compared with the depth. The hyperbolic tangent term becomes 1, so that the velocity is independent of depth. For $\lambda = 1$ cm (and $D > \frac{1}{6}$ cm),

$$v = \sqrt{\left(9.8 \times \frac{0.01}{2\pi} + \frac{73 \times 10^{-3}}{(1 \times 10^{3})(0.01/2\pi)}\right)}$$

$$= \sqrt{(1.6 \times 10^{-2}) + (4.6 \times 10^{-2})} = 0.25\text{m/s}.$$

This is in the region of ripple tank waves, or of ripples in a pond where the wavelength might be a foot or two, with the depth being larger than a few inches. Note that the effect of surface tension is three times greater than the effect of gravity. However, if $\lambda = 3$ cm, with $D/\bar{\lambda} > 1$, the surface tension effect decreases by 3 and the gravitational

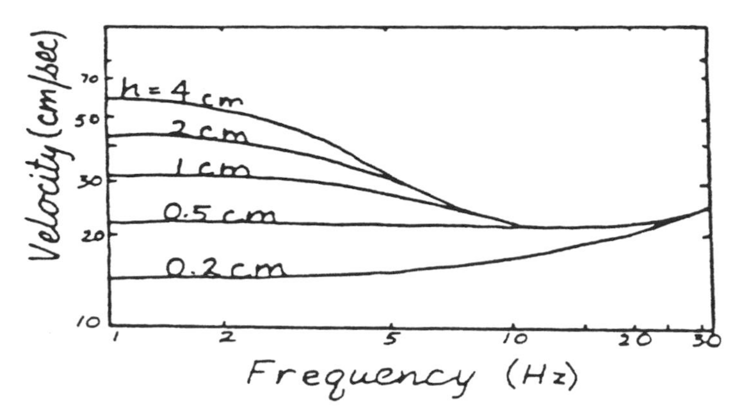


FIGURE 3.1 Velocity of water ripples as function of frequency and water depth.

effect increases by 3. Note that although the velocity does not depend on depth, it does depend on $\lambda$. The medium is *dispersive*: a wave that has components of various wavelengths will spread out, for the components travel with different velocities.

Now consider cases where $D/\bar{\lambda} < 1$, and therefore $\tanh(D/\bar{\lambda}) \approx (D/\bar{\lambda})$. In a ripple tank the relationships linking depth and wavelength to velocity are complex and are best understood in terms of a graph, as shown in figure 3.1. To demonstrate refraction in a ripple tank, choose a frequency less than 5 Hz and have a depth of only a few mm in one part and over 1 cm in the other part. For non-dispersion, keep the depth uniform at about 5 mm with a frequency greater than 5 Hz. For lake and ocean waves, the amplitudes are large and the surface tension effect is negligible. What remains is

$$v = \sqrt{(g\bar{\lambda})(D/\bar{\lambda})} = \sqrt{gD}.$$

Under these conditions, the medium is not dispersive. This is the equation for the velocity of the actual tidal waves, or of tsunamis in deep oceans. For instance, in the Pacific a tsunami travels with speed

$$v = \sqrt{(9.8)(4 \times 10^3)} = 200 \text{ m/s} \approx 440 \text{ mi/hr}.$$

In Long Island Sound, where the depth is about 30 m, the velocity is 17 m/s. The Sound acts like a long rectangular trough, closed at one end. The wavelength of the fundamental vibration of such a trough or tube is four times the length, which in this case is about 140 km. The period is

$$T = \frac{\lambda}{v} = \frac{5.6 \times 10^5 \text{ m}}{17 \text{ m/s}} = 3.3 \times 10^4 \text{ s} = 9 \text{ hr}.$$

The semiperiod of the moon is 12½ hr. This is close enough to the natural period of the Sound so that a sizable resonance is produced with a Q of about 3. The mean tidal range in height at the open (eastern) end is about 2 ft, and at the closed (western) end over 6 ft.

## COLD FLAT MUSIC

The velocity of sound in air depends on the temperature. An approximate formula is

$$v = (331 + 0.6t) \text{ m/s}.$$

The parameter, $t$, is the temperature in degrees Celsius. At 20°C, for example, $v = 343$ m/s.

What happens to the pitch produced by wind instruments as they get colder? If they are tuned in a warm room and then go out for a winter band parade, the resonant wavelengths in the horns will remain the same, but the frequency will decrease. (The wavelengths are detemined by the physical size of the instrument, which will not change appreciably.) Let's calculate the effect and see if it is large enough to require retuning:

$$\lambda f = v \quad \Rightarrow \quad \frac{\Delta f}{f} = \frac{\Delta v}{v} = \frac{12}{343} = 3.5\%.$$

That percentage change in frequency would lower the middle A to which the orchestra tunes from 440 Hz to 425 Hz. That takes the pitch down about one-third the way to G, an effect that is clearly noticeable.

## THUNDER AND LIGHTNING

Everyone knows that you see the lightning before you hear the thunder. The arithmetic for the time delay is particularly simple if you use old-fashioned units. The speed of sound in air at 20°C is 343 m/s, which equals 1130 ft/s. In 5 s, the sound will travel 5650 ft, which is remarkably close to the 5280 feet in a mile. Most thunder storms occur in the summertime when it is hotter than 20°C and the speed of sound is faster. An approximate formula for the speed of sound as a function of temperature is

$$v = (331 + 0.6t)\ \mathrm{m/s}, \text{ where } t \text{ is Celsius temperature.}$$

At 30°C, $v = 349\ \mathrm{m/s} = 1150$ ft/s.

As a practical matter, the usual timing of the delay involves errors greater than those introduced by the rule of allowing one second for a fifth of a mile. When you see the lightning, start counting "one thousand and one, one thousand and two," etc. Each syllable takes one fifth of a second. If your count is less than "one thousand and two" you should not be under a tall tree.

## WAVELENGTHS OF FAMILIAR SOUNDS

The effects of different audio wavelengths can easily be heard by listening to a high-fidelity speaker. Diffraction shadows will be produced by people or stuffed furniture that are larger than the audio wavelengths. Long wavelength notes will curl around the furniture, around corners, and down the hall. Let's calculate the wavelengths of some familiar sounds.

The speed of sound in air at room temperature is 344 m/s. The orchestra tunes to the sound of the oboe playing middle A, with a frequency of 440 Hz. Since $\lambda f = v$, the wavelength of the middle A note is

$$\lambda = \frac{344 \text{ m/s}}{440 \text{ Hz}} = 0.78 \text{ m} \approx 30 \text{ in.}$$

The lowest note from orchestral instruments comes from the bass viol at 39 Hz. This is about the limit of human hearing as a musical tone. That wavelength is

$$\lambda = \frac{344 \text{ m/s}}{39 \text{ Hz}} = 8.8 \text{ m} \approx 29 \text{ ft.}$$

At the other extreme, the violin can reach high E at 2637 Hz. That wavelength is

$$\lambda = \frac{344 \text{ m/s}}{2637 \text{ Hz}} = 13 \text{ cm} \approx 5 \text{ in.}$$

The piano can reach slightly beyond each of these extremes, from 12.5 m to 8 cm. So, low notes get diffracted by a couch; high notes by a martini glass. (Fig. 3.2.)

The lowest note that a musical instrument can sound is linked to its size. As a rule of thumb, an instrument does not produce a musical sound with a wavelength in air much longer than the instrument itself. In the case of a flute, the lowest note has a wavelength about twice the length of the flute. In the case of the trumpet, the lowest note that can be sounded is about the length of the instrument (about 5 ft). With the violin, the lowest note produced by the G string is about four times the length of the case.

Some bats use a form of audio radar to catch insects. The wavelength must be as small as, or smaller, than the prey. Otherwise, the

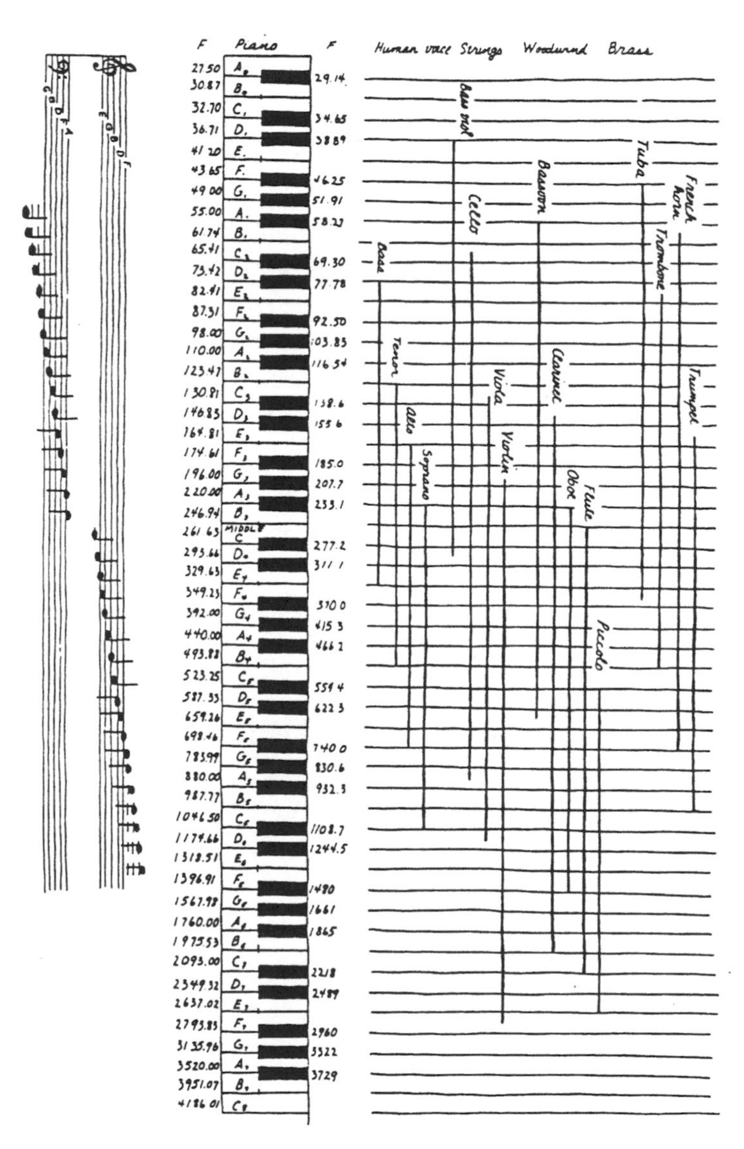


FIGURE 3.2 Piano keyboard relationships.

waves diffract around the prey, leaving very little energy to be reflected. Assume that $\lambda = 5$ mm. Then

$$f = \frac{344 \text{ m/s}}{5 \times 10^{-3} \text{ m}} = 70,000 \text{ Hz}.$$

This frequency is in the supersonic range, far above human hearing.

Porpoises also use sonic radar, with a chirping frequency span from 50,000 to 100,000 Hz. Their prey is small fish, perhaps 10 cm long. It may seem strange that the frequency range for porpoises would be the same as that of bats, who are after much smaller food. However, the speed of sound in water is greater than that in air by a factor of almost 5.

$$\lambda = \frac{1530 \text{ m/s}}{5 \times 10^{4} \text{ Hz}} = 0.03 \text{ m} = 3 \text{ cm}.$$

It seems that many creatures survive by knowing physics.

## SCALES AND CHORDS

Surprisingly, musical harmony is based on arithmetic. The overtones produced by most musical instruments are harmonic—that is to say, the overtones are integral multiples of the fundamental frequency. Combinations of notes sound harmonious or cacophonous to us, depending on our cultural background and on whether the mixture produces beats. Our standard music scales are based on numerical permutations of the frequencies of certain chords.

**Table 3.1**
**The Scale of C**

| *C* | *D* | *E* | *F* | *G* | *A* | *B* | *C* | *D* |
|---|---|---|---|---|---|---|---|---|
| do | re | mi | fa | sol | la | ti | do | re |
| 264 | 297 | 330 | 352 | 396 | 440 | 495 | 528 | 594 |

**Table 3.2**
**A Scale Constructed with Equal Frequency Increments**

| 264 | 302 | 339 | 377 | 415 | 453 | 490 | 528 | 566 |
|---|---|---|---|---|---|---|---|---|
| C | D | E-F | F# | G# | A-A# | B | C | C# |
| White | Wh | Wh | Bl | Bl | Wh | Wh | Wh | Bl |

Take, for instance, the standard scale of C, the one that can be played with all white keys on the piano, starting with middle C at 264 Hz (cycles per second). This scale is based on a chord called the *major triad.* The three notes of the triad should have frequencies in the ratios of 4:5:6. These ratios generate the frequencies 264:330:396. The second triad consists of 352:440:528, the third consists of 396:495:594. This is the do, re, mi scale in the key of C. The names and frequencies of the notes are shown in table 3.1.

Note that 330/264 = 5/4, 396/330 = 6/5, and 396/264 = 6/4. The note that the orchestra tunes to is middle A at 440 Hz. An increase of an octave corresponds to a doubling of frequency. Note the octave jumps with C and D. The re at 297 is an octave below the re at 594.

There are eight white piano keys in an octave. Why not just divide the frequency spread of an octave into seven evenly spaced steps? Dividing 264 by 7 yields frequency increments of 37.7. The scale and piano notes would then be as shown in table 3.2.

Try playing that scale on a piano, singing do, re, mi . . . . Use the sequence of white and black keys shown.

Instead of *adding* equal frequency increments, let's add equal percentage increments. Multiply the frequency of each note by $(1 + \varepsilon)$ to generate the next note. If we do that seven times starting with C, we must get back to the next higher C, one octave up.

$$(1 + \varepsilon)^7 = 2; \quad 7 \log(1 + \varepsilon) = \log 2 = 0.30;$$

$$\log(1 + \varepsilon) = 0.043; \quad 1 + e = 1.104$$

Compare the scale generated this way with the "perfect key of C" shown in table 3.3.

**Table 3.3**
**A Scale Constructed with Equal Percentage Increments**

| | | | | | | | | |
|---|---|---|---|---|---|---|---|---|
| Key of C | 264 | 297 | 330 | 352 | 396 | 440 | 495 | 528 |
| % increment | 264 | 291 | 322 | 355 | 392 | 433 | 478 | 528 |

**Table 3.4**
**The Equal-Tempered Scale**

| *Note* | *C C#* | *D D#* | *E* | *F F#* | *G G#* | *A A#* | *B* | *C* |
|---|---|---|---|---|---|---|---|---|
| Name | do | re | mi | fa | sol | la | ti | do |
| Frequency of even-tempered scale | 264 | 296.3 | 332.6 | 352.4 | 395.6 | 444 | 498.4 | 528 |
| Frequency of perfect key of C | 264 | 297 | 330 | 352 | 396 | 440 | 495 | 528 |

We can come even closer by including the black piano keys and using a multiplying increment of 1.05946, derived from the arithmetic for twelve keys, This is called the *equal-tempered scale* (table 3.4). See how well it matches the "perfect key of C."

Of course, musicians have their own language to describe all this—octave, triad, diminished seventh, augmented fifth, major and minor chords. In the making of music, their language is more convenient and perhaps more powerful than that of the physicist.

## THE SOUND OF COKE

If you can find a Classic Coke™ bottle, blow across the top to produce the resonant "lip" tone. (Or use any bottle.) You can find the frequency of the note by matching it against a piano keyboard. For the Coke bottle, we found a reasonable match with B flat, below middle C, at a frequency of 233 Hz.

Suppose we model the bottle as a cylindrical tube closed at one end. Then the wavelength should be four times the length of the tube. The inside height of the Coke™ bottle is about 18.5 cm. Therefore, the wavelength would be 74 cm. The frequency would be

$$f = \frac{v}{\lambda} = \frac{343 \text{ m/s}}{0.74 \text{ m}} = 464 \text{ Hz}.$$

Clearly, we have chosen a poor model! Most soft-drink bottles are not cylindrical tubes, but have various curved shapes that include a tapering neck. A better model would be a Helmholtz oscillator, which is shown in figure 3.3.

The crucial parameters of the oscillator are the cross-sectional area of the neck ($A$); the length of the neck ($\ell$); and the volume of the bulb ($V$). The model assumes that a pulse of air is sent down the neck, compresses the air in the bulb, and bounces back up again. The resonant frequency is

$$f = \frac{v_{\text{sound}}}{2\pi}\sqrt{\frac{A}{V\,\ell}}.$$

The values of these parameters for a Classic Coke™ bottle are $A = 2.4 \times 10^{-4}$ m$^2$; $V = 2.25 \times 10^{-4}$ m$^3$; $\ell = 4.45 \times 10^{-2}$; $v_{\text{sound}} = 343$ m/s. The measurements involve subjective judgments about where the neck begins and ends. With these reasonable values, the frequency is

$$f = \frac{343}{2\pi}\sqrt{\frac{2.4 \times 10^{-4}}{(2.25 \times 10^{-4})(4.45 \times 10^{-2})}}$$

$$= 267 \text{ Hz (about middle C)}.$$

Evidently a Classic Coke™ bottle is not a pure Helmholtz oscillator. The neck does not have uniform diameter, and its volume merges into that of the bulb. But it's a better model than that of a uniform diameter

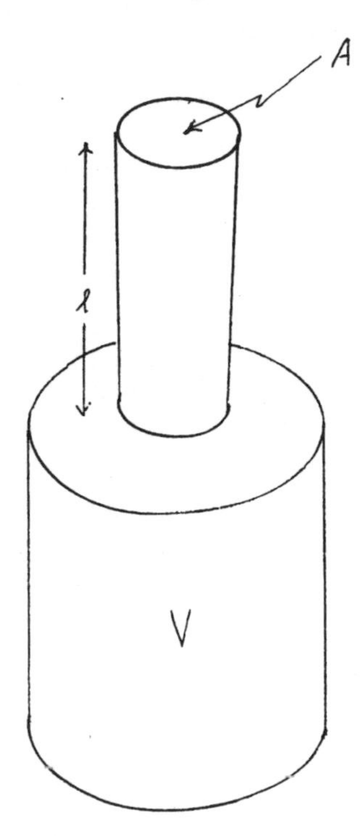


FIGURE 3.3 Helmholtz resonator geometry.

tube, and the model furnishes a guide to approximating the resonant behavior of other bottles.

A more detailed analysis was given in an article in *The Physics Teacher* 36 (February 1998): 70, by Mark Silverman and Elizabeth Worthy.

## THE SENSITIVE EAR

The sensitivity of the human ear is astonishing. Like most of the sense organs of the body, the ear responds logarithmically. For every multiplicative factor of 10 to the input, the sensor responds by an arithmetic change of one unit. Thus an increase of intensity of sound from $10^{-11}$ to $10^{-10}$ W/m$^2$ is heard as an increase from 10 to 20 decibels. (The unit is 10 db, rather than 1 bel.) If the sound intensity increases from $10^{-10}$ to $10^{-9}$, the ear hears an increase from 20 to 30 db.

At the frequency of 1000 Hz to which the ear is most sensitive (two octaves above middle C), the weakest signal we can hear has an intensity of about $10^{-12}$ W/m$^2$, corresponding to an excess air pressure of $3 \times 10^{-5}$ N/m$^2$. This lower limit of intensity was determined experimentally, and rather arbitrarily, with human testers. The value of $10^{-12}$ then becomes the reference origin for the decibel scale of loudness.

The excess pressure, $p$, produced by a sound wave depends on the bulk modulus, $B$, of the gas, the wavelength, $\lambda$, and the amplitude, $A$, of the sinusoidal oscillation:

$$p = B\frac{2\pi}{\lambda}A.$$

The displacement amplitude is

$$A = \frac{p}{B}\left(\frac{\lambda}{2\pi}\right) = \frac{(3 \times 10^{-5}\ \mathrm{N/m^2})}{(1.4 \times 10^5\ \mathrm{N/M^2})}\left(\frac{0.3}{2\pi}\ m\right) = 1 \times 10^{-11}\ \mathrm{m}.$$

This is ten times smaller than the radius of an atom! Nevertheless, our ears can respond to such a weak signal.

Incidentally, the bulk modulus for air is equal to $\gamma P_{\mathrm{atm}}$, where $\gamma$ is the ratio of specific heats $c_P/c_V$ and $P_{\mathrm{atm}}$ is atmospheric pressure. For air, $\gamma = 1.4$.

## λ OF VISIBLE LIGHT

Here's how you can calculate the wavelength of visible light without making a measurement.

The smallest detail that you can see with the naked eye has a size of about 1 mm. The greatest magnification of an optical microscope is about 1000. (To be sure, with oil immersion—between objective lens and the object—using blue light, the magnification can be increased to about 2000.) Therefore, the smallest detail that you can see with a microscope is about $10^{-6}$ m. (A magnification of 1000 would make an object of $10^{-6}$ m look like 1 mm.)

Why not buy a more "powerful" and expensive microscope? It doesn't exist. Diffraction of light obscures details that are smaller than $10^{-6}$ m. Therefore, the wavelength of visible light must be of the order of $10^{-6}$ m. Indeed, $\lambda_{\text{red}} = 0.7 \times 10^{-6}$ m and $\lambda_{\text{blue}} = 0.4 \times 10^{-6}$ m.

$$\lambda\,\nu = c; \qquad h\,\nu = E$$

Sometimes we know or can estimate the wavelength of some particular electromagnetic radiation, and so we can calculate its frequency—or vice versa. Here are some familiar examples.

The wavelength of visible light must be a fraction of a micron, about $\frac{1}{2} \times 10^{-7}$ m (see the previous section). Therefore, the frequency of visible light must be

$$\nu = \frac{3 \times 10^{8}\ \text{m/s}}{\frac{1}{2} \times 10^{-6}\ \text{m}} = 6 \times 10^{14}\ \text{Hz}.$$

The energy of each photon is

$$h\,\nu = (6.6 \times 10^{-34}\ \text{J} \cdot \text{s})(6 \times 10^{14}\ \text{Hz})$$
$$= 4.0 \times 10^{-19}\ \text{J} = 2.5\ \text{eV}.$$

This is enough energy to break up certain molecules, causing bleaching, tanning, or production of photographic images.

The maximum voltage of a dentist X-ray machine is $10^5$ volts. Thus, the energy of the X-ray photon is $10^5$ eV, and its frequency is

$$\nu = \frac{(1 \times 10^{5}\ \text{eV})(1.6 \times 10^{-19}\ \text{J/eV})}{6.6 \times 0^{-34}\ \text{J} \cdot \text{s}} = 2.4 \times 10^{19}\ \text{Hz}.$$

The wavelength of this photon is $\lambda = \frac{3\times10^{8}}{2.4\times10^{19}} = 1.3 \times 10^{-11}$ m. This is about one-tenth the size of an atom.

A police radar detector is about the size of a large flashlight. The diffraction limited angle of its beam is $\theta = \frac{\lambda}{d}$, where $d$ is the diameter of the "gun," about 10 cm. To separate cars at a reasonable distance, $\theta$ must be no larger than $10° = 0.18$ radians. Therefore,

$$\lambda = (0.18 \text{ rad})(0.10 \text{ m}) = 0.02 \text{ m} = 2 \text{ cm}.$$

The corresponding frequency is $\frac{3\times10^{8}}{0.02} = 15 \times 10^{9}$ Hz, which is in the microwave region. The energy of the 2 cm photon is

$$\begin{aligned} h\,\nu &= (6.6 \times 10^{-34} \text{ J} \cdot \text{s})(1.5 \times 10^{10} \text{ Hz}) \\ &= 9.9 \times 10^{-24} \text{ J} = 6.2 \times 10^{-5} \text{ eV}. \end{aligned}$$

This energy is well below the energy of molecular binding, or average vibrational or translational kinetic energy of molecules at room temperature. It is, however, in the energy range of molecular *rotations* and so can be used in the study of complex molecules. Wavelengths below 1 cm are strongly absorbed by water moisture in air.

The home microwave oven has a frequency of 2450 MHz, with a wavelength of $\frac{(3\times10^{8} \text{ m/s})}{2.45\times10^{9} \text{ Hz}} = 12$ cm. This wavelength is conveneiently small in comparison with the home oven but is large enough so that penetration depths of a couple of wavelengths can cook most pieces of food. (The heating is caused by dielectric vibrations; there is no effect of a water resonance.)

Before the days of cable TV, every house roof had an antenna aimed at the TV stations. The length of the bars on these antennas was a clue to the frequencies of the stations. The easiest one to figure out was the Yagi antenna, which has a row of rods parallel to each other. Each rod served as a folded dipole, or reflector thereof, and was about $\frac{1}{2}\lambda$ long. A typical Yagi had rods about 1.5 m long. A wavelength of 3 m corresponds to a frequency of $\frac{(3\times10^{8} \text{ m/s})}{3 \text{ m}} = 100$ MHz. This is in the middle of the FM stations, which are between TV channels 6 and 7.

In the case of AM radio, we know the frequency but seldom worry about the wavelength and are never concerned with the corresponding photon energy. The AM band in the United States goes from 535 kHz to 1605 kHz. Taking the middle at 1 MHz as a representative frequency, $\lambda = \frac{3\times10^8}{1\times10^6} = 300$ m. It's no wonder that radiowaves can diffract around trees and houses. A dipole antenna for this wavelength requires a high tower, but it doesn't have to be half a wavelength high. The land underneath the tower is covered with radial conductors, forming a mirror for the radio waves. A tower that is 75 m tall appears with its reflection to be 150 m tall. Of course, the *receiving* antenna need not be half a wavelength long. The transmitted field is measured in volts per meter. The more meters in the (vertical) receiver antenna, the more volts on the receiver input. As a practical matter, the antenna may be only a few centimeters high. Note, however, that the E field is vertical since the transmitter antenna is vertical.

## TRAINS AND DOPPLER SHIFT

In descriptions of the Doppler shift many decades ago, an appeal was generally made to the common experience of hearing a train whistle. If you are beside the railroad crossing as the train is hurtling toward you, you hear a lonesome whistle calling. The whistle gets louder as the train draws near, but its pitch does not change. Just as the train passes you, the pitch abruptly drops. It's the Doppler effect, but the example is not so useful anymore. Many students have never heard a train whistle at a railraod crossing.

Let's put in numbers to calculate the change of pitch. The train is traveling at $v = 60$ mph (27 m/s); the speed of sound, $V$, is 340 m/s; and the frequency of the whistle at rest is 1000 Hz:

$$f_{\text{observed}} = f_{\text{source}} \left( \frac{V}{V \pm v} \right)$$

$$= f_{\text{source}} \left( \frac{1}{1 \pm v/V} \right) \approx f_{\text{source}}(1 \mp v/V).$$

In the final expression, the + sign is for the approaching train; the – sign is for the retreating train. The approximation is valid, since $v/V$ is << 1.

$$f_{\text{observed}} = 1000 \text{ Hz } (1 \mp 0.079).$$

The frequency is 8% higher than 1000 Hz for the approaching train, and 8% lower when it is retreating. In music terminology, the basic frequency of 1000 Hz is a little higher than high B (an octave above middle B on the piano). The approaching whistle sounds like high C-sharp, and the departing train sounds like a B-flat. This abrupt shift of pitch is dramatic and easily heard.

# Chapter 4

# Heat

## HUMAN HEATERS

If you have ever been in a crowded room on a hot day, you know that humans generate heat. But how much?

The heat output must be equal to the energy input. A starvation diet for an average human is about 1500 diet calories per day. That's enough to stay alive but not enough to allow one to produce much physical work. The calories going in must equal the heat going out:

$$1500\ \frac{\text{kcal}}{\text{day}} \times \frac{10^3\ \text{cal}}{1\ \text{kcal}} \times \frac{4.2\ \text{J}}{\text{cal}} \times \frac{1\ \text{hr}}{3600\ \text{s}} \times \frac{1\ \text{day}}{24\ \text{hr}} = 73\ \text{W} \approx 75\ \text{W}.$$

Get a dozen or so people in a room, and it's the equivalent of having a one-kilowatt space heater with you.

## NEGATIVE CALORIE DIET

Picture a person sitting down at a table filled with high-calorie foods. There is butter in the gravy on the mashed potatoes. There is whipped cream on the icing on the cake. And there is one other thing on the table: ice water. Calories are calories, although the diet calorie is actually a kilocalorie. For each cubic centimeter (one gram) of ice water that you ingest, the body must supply 37 of the little calories to raise the temperature of the water to body temperature at 37°C. How much ice water must you drink to cancel a daily intake of 3000 diet calories?

$$(37 \text{ cal/cc})(\#\text{cc}) = (3000 \times 10^3 \text{ cal});$$
$$\#cc = 81 \times 10^3 \text{ cc} = 81 \ \ell \approx 20 \text{ gal}.$$

It doesn't make any difference whether you drink it or sit in it all day long—you'll burn the calories either way. Of course, you can do even better by chewing ice cubes. It takes 80 calories to melt each gram of ice.

## THE MORAL OF THE TAIL

Isn't it amazing that at 300 K you and a Christmas goose are chemically quite stable, but at only twice that temperature, at 600 K, you and your goose would be rapidly cooked? The threshold energy for a chemical transition is about 1 eV. The average kinetic energy of a molecule at room temperature (300 K) is about $\frac{1}{25}$ eV and at 600 K would be about $\frac{1}{12}$. (See the section "Units and Approximations" at the beginning of the book.) It would seem as if you and the goose are well below the threshold for cooking.

Consider, however, the Maxwell-Boltzmann curve of kinetic energy distribution. The formula for the curve is

$$\Delta N = \frac{2\pi}{(\pi k T)^{\frac{3}{2}}} N \, e^{-\frac{E}{kT}} \sqrt{E} \, \Delta E \ ,$$

where $N$ is the total number of particles, $k$ is Boltzmann's constant $= 1.38 \times 10^{-23}$ J/mol · K, and $\Delta N$ is the number of molecules in the energy range $\Delta E$.

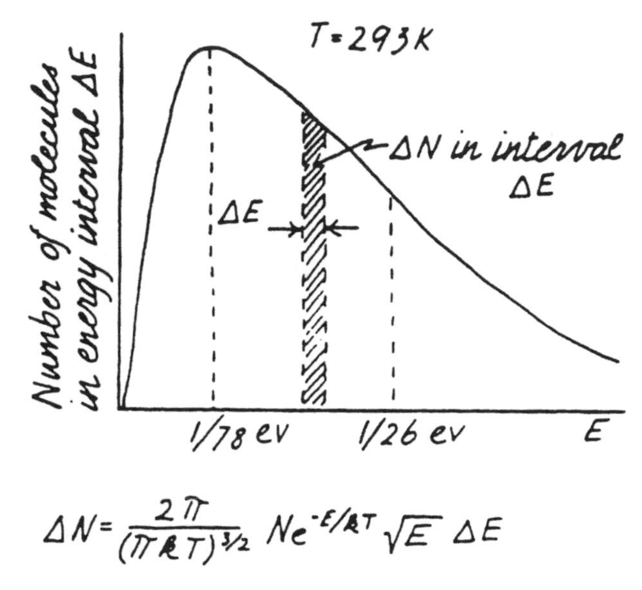


FIGURE 4.1 Maxwell-Boltzmann energy distribution of molecules.

Note that for $E$ small compared with $kT$, the value of the exponential is about 1. The rise of the curve from $E = 0$ is controlled by the factor $\sqrt{E}$. For $E > kT$, the exponential dominates. Figure 4.1 shows the number of atoms in the energy range $\Delta E$ as a function of $E$. The average energy of $\frac{1}{25}$ eV is well below the threshold energy of 1 eV, shown here necessarily out of scale. The question is: Out of $N$ atoms, how many have energy above $E_{\text{threshold}}$? These are the ones making the chemical transformations that define cooking. To find out, integrate the curve from $E = E_{\text{threshold}}$ to $E = \infty$. A good approximation is that $\sqrt{E}$ stays constant and $\frac{2}{\sqrt{\pi}} \approx 1$.

$$N_{>E_{\text{threshold}}} = N\sqrt{\frac{E_t}{kT}}\, e^{-\frac{E_t}{kT}} .$$

For $T_{\text{room}}$, $kT = \frac{1}{40}$. Therefore,

$$N_{>1\text{ eV}} = N\sqrt{40}\, e^{-40} = 3 \times 10^{-17} N.$$

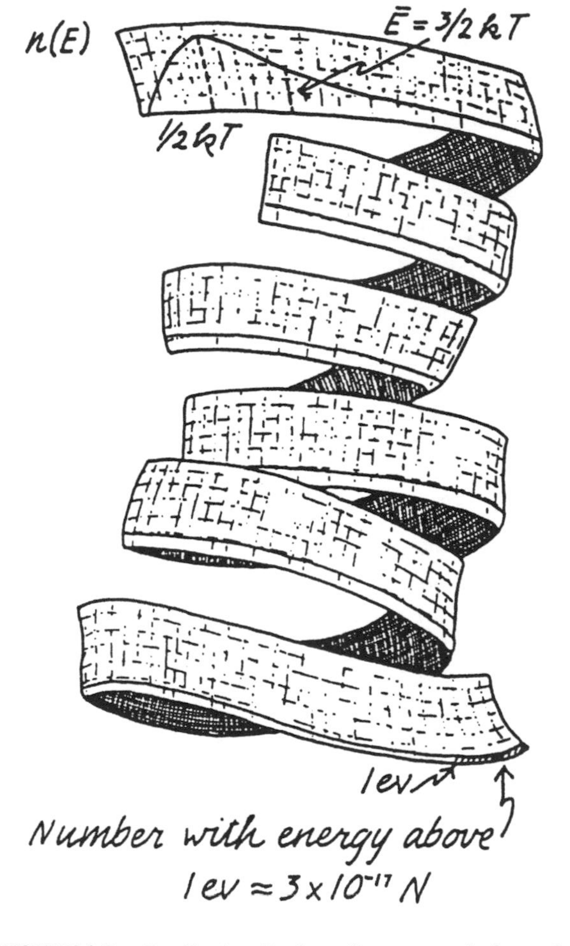


FIGURE 4.2 Qualitative display of energy scale for molecules, showing huge range between average energy and energy needed to cause chemical change.

For $N = 1\text{mole} = 6 \times 10^{23}$, $N_{>1\,\text{eV}} = 2 \times 10^{7}$. Twenty million chemically active particles seems like a lot, but consider the situation at twice the temperature. At $T = 600$ K, $E_{\text{ave}} = \frac{1}{20}$ eV.

$$N_{>1\,\text{eV}} = N\sqrt{20}\,\text{e}^{-20} = 1 \times 10^{-8}\,N.$$

By doubling the temperature from 300 K to 600 K, the number of atoms over the threshold energy is greater by a factor of $\frac{1\times10^{-8}}{3\times10^{-17}} = 3 \times 10^{8}$.

In chemical interactions, the action is in the tail of the energy distribution, and the number in the tail is very sensitive to the temperature and the threshold energy. (Note that the arithmetic would be the same whether we doubled the temperature or lowered the threshold energy by a factor of 2.) Figure 4.2 shows the scale of the exponential dependence.

## ICE SKATING

The low friction of ice skating is caused by a thin film of water between ice and skate blade. The popular explanation for the existence of the water is that the pressure of the blade on the ice lowers the melting point. The evidence cited for this effect comes from the phase diagram of water (figure 4.3).

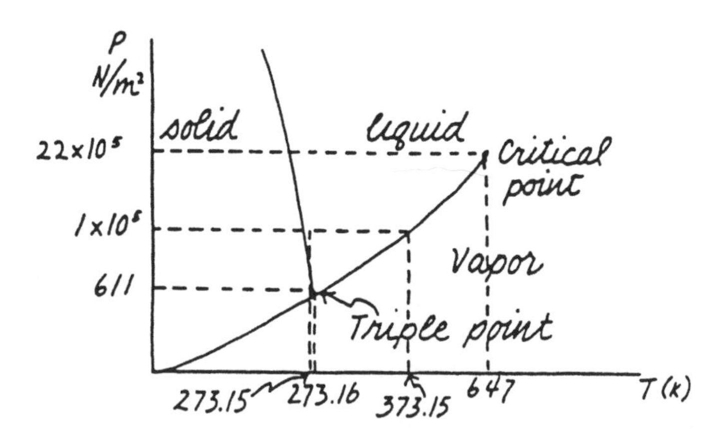


FIGURE 4.3 Common pressure-temperature phase diagram for water.

At the atmospheric pressure of $1 \times 10^5$ N/m$^2$, the melting point of ice is 0°C. The boundary curve between solid and liquid has a negative slope. As the pressure increases, the melting point decreases. Hence the argument that the pressure of the blade melts the ice, creating the low friction of a water film. The reasoning is a splendid example of the need for quantitative treatment of theories. *Put in the numbers!* The slope of that phase curve between solid and liquid is $-1.2 \times 10^7$ (N/m$^2$)/ °C $= -120$ atm/°C. If the graph of $P$ versus $T$ were drawn to scale, the slope line between solid and liquid would be vertical.

A typical skate blade is 27 cm long by 4 mm wide. That's an area of 11 cm$^2$. If a 65 kg person stands on one skate, the pressure is $(650\ \text{N})/(11 \times 10^{-4}\ \text{m}^2) = 6.5 \times 10^5\ \text{N/m}^2 \approx 6$ atm. That would lower the melting temperature about $1/20°$.

One might argue that the blades are sharpened, thus exerting greater pressure along the edges. The sharp edge simply sinks into the ice, leaving the whole area for support. Perhaps the film of water is caused by the friction of the front edge of the blade striking the ice. But that doesn't explain why the sliding friction is also low for hockey pucks.

About 170 years ago, Faraday proposed that there might be a transition layer between ice and water. He didn't have our modern concept of molecules, but we know now that such a layer exists and is highly temperature sensitive. At 0°C the layer of water is several hundred molecules thick, and at $-10$°C it is almost gone. One might think, therefore, that the lowest friction and the fastest skating would occur when the ice is at 0°C. There is another factor, however, and that is the softness of the ice. At 0°C, the blade crushes the ice and sinks in, increasing friction. At $-10$°C, the ice is hard, but there is no layer of water and the skating is sticky. The optimum condition for speed skating occurs at $-7$°C.

For further details, see the article by J. D. White, "The Role of Surface Melting in Ice Skating," *The Physics Teacher* 38 (1992): 495.

# THERMAL EXPANSION

Why do solids expand as they get hotter? It might seem that it is because the atoms in their potential wells oscillate with greater amplitudes as thermal energy is fed in. Consider, however, that if the potential wells were symmetric, the equilibrium position of each atom would not change with temperature. At the outer faces of a solid, the greater amplitude of oscillation would expand the domain of the surface atoms, but by a trivial amount. Instead, thermal expansion occurs because the potential well is asymmetric, as shown in figure 4.4. As the temperature rises and the kinetic energy of each atom increases, the amplitude of oscillation increases. That forces an increase in the equilibrium position.

Let's put numbers in and calculate the increase in oscillation amplitude and the increase in the equilibrium position.

For oscillation of an atom in a metal,

$$U_{\text{thermal}} = 3\,kT \quad (\frac{3}{2}\,kT \text{ kinetic, } \frac{3}{2}\,kT \text{ potential}).$$

At 300 K,

$$U_{\text{thermal}} = 3(1.38 \times 10^{-23}\ \text{J/atom} \cdot \text{K})(300\text{K})$$

$$= 1.2 \times 10^{-20}\ \text{J/atom} = \frac{1}{13}\ \text{eV/atom} \Rightarrow \frac{1}{26}\ \text{eV/atom, oscillation.}$$

The oscillation energy is also given by $\frac{1}{2}\kappa A^2$, where $\kappa$ is the atomic "spring constant" and $A$ is the amplitude of oscillation:

$$\frac{1}{2}\kappa A^2 = 1.2 \times 10^{-20}\ \text{J/atom}.$$

The value of $\kappa$ is 50 J/m$^2$. (See the section "Atomic Spring Constant" in chapter 9.)

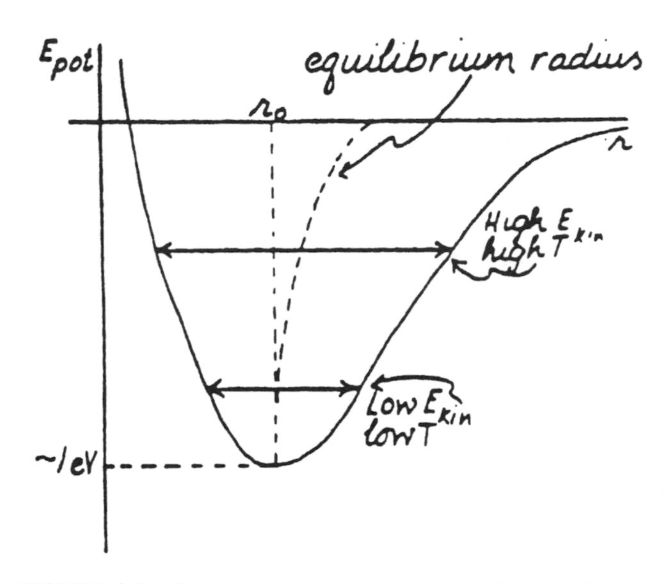


FIGURE 4.4 Asymmetric molecular potential well, showing shift of equilibrium radius as temperature increases.

$$A = \sqrt{\frac{2\,(1.2 \times 10^{-20})}{50}} = 2.2 \times 10^{-11}\ \text{m} = 0.22\ \text{Å}.$$

If $T$ doubles, $A$ increases by a factor of 1.4. For 600 K, $A = 3.1 \times 10^{-11}$ m, and $\Delta A \approx 0.09$ Å. The increase in the equilibrium radius (which causes expansion) can be found from the thermal expansion coefficient,

$$\frac{\Delta V}{V} = \text{B}\,\Delta T.$$

A typical value of B for metals is $20 \times 10^{-6}\ \text{K}^{-1}$:

$$\frac{\Delta V}{V} = (20 \times 10^{-6})\,300 = 6 \times 10^{-3},$$

$$\frac{\Delta L}{L} = \frac{1}{3}\frac{\Delta V}{V} = 2 \times 10^{-3}.$$

For an atomic radius, $L = 1 \times 10^{-10}$ m, and $\Delta L = 2 \times 10^{-13}$ m = 0.002 Å.

Note that the increase in equilibrium position due to the asymmetry of the potential well is only about one-fortieth the increase in the amplitude of oscillation. But it is the increase in equilibrium position that causes the solid to expand.

Note also that the work involved in expanding the solid is small compared to the thermal energy fed in and stored in the internal energy form. When the solid's temperature is increased by 300 K, the added energy per atom is

$$\frac{(1.2 \times 10^{-20}\ \mathrm{J})}{(1.6 \times 10^{-19}\ \mathrm{J/eV})} = 7.5 \times 10^{-2}\ \mathrm{eV} = \frac{1}{13}\ \mathrm{eV}.$$

Stretching the solid (in one dimension) by the amount it expands when heated by 300 K requires

$$\frac{1}{2}(50\ \mathrm{J/m^2})\,(2 \times 10^{-13}\ \mathrm{m})^2 = 1.0 \times 10^{-24}\ \mathrm{J} = 6.3 \times 10^{-6}\ \mathrm{eV}.$$

The first law of thermodynamics is: $Q = \Delta U + P\,\Delta V$. In the case of a solid, both the kinetic energy of oscillation and the spring-like energy of stretching of the equilibrium position are part of the internal energy, $U$. The product, $P\Delta V$, accounts for external work done against an outside pressure as the solid expands. For a solid expanding against atmospheric pressure, the $P\Delta V$ term is negligible compared with the increase in $U$. This internal energy goes into thermal oscillation of the atoms, and a small amount goes into an increase in their equilibrium position.

## POWER PLANT EFFICIENCY

You can't beat Carnot's efficiency predictions. You can turn mechanical or electrical energy into heat with 100% efficiency, but you can't do the reverse using a cyclic machine without throwing away some energy. The first law of thermodynamics ($Q = \Delta U + P\,\mathrm{dV}$) says that you can't get something for nothing. The second law ($\Delta S \geq 0$) says that you can't even break even.

How well do the big electric generating plants do? They certainly throw away a lot of energy by heating up rivers or the air, or discharging water at lower levels. The Carnot condition in terms of boiler temperature, $T_1$, and discard temperature, $T_2$, is

$$\text{efficiency, } \varepsilon = \frac{T_1 - T_2}{T_1}.$$

This formula for the ideal efficiency of a heat engine operating between two temperatures is not a good approximation for actual working conditions. The formula assumes that the working fluid extracts heat from the boiler at constant temperature $T_1$, and then discharges it at constant temperture, $T_2$. But if much energy is to be sent from the boiler to the working substance, there must be a considerable temperature difference to make the heat flow. A more realistic efficiency formula is

$$\varepsilon = \frac{\sqrt{T_1} - \sqrt{T_2}}{\sqrt{T_1}}.$$

This formula is described and justified in an article by F. L. Curson and B. Ahlborn, "Efficiency of a Carnot Engine at Maximum Power Output" *Am. J. Phys.* 43 (1975): 22.

The Carnot engine efficiency of a steam plant operating between 873 K and 293 K (600°C and 20°C) is

$$\varepsilon_{\mathrm{Carnot}} = \frac{873\ \mathrm{K} - 293\ \mathrm{K}}{873\ \mathrm{K}} = 66\%.$$

The Curson-Ahlborn efficiency is

$$\varepsilon_{\mathrm{C-A}} = \frac{\sqrt{873} - \sqrt{293}}{\sqrt{873}} = 42\%.$$

This is the efficiency approximately obtained by commercial plants. The lower temperature is determined by the temperature of the “sink” into which they can pour the cooling water. If the sink is a river or lake, the efficiency increases slightly in the winter. If the operating temperature is increased much beyond 873 K, the water will start dissociating, and the freed oxygen will eat up the boiler.

*Nuclear* power plants operate at lower boiler temperature, lest the hot working fluid weaken the cladding on the uranium fuel rods. A typical operating temperature for a nuclear plant is 600 K, leading to an efficiency of 30%.

Note that in a fossil fuel burning plant, for every joule of heat produced in the boiler, 0.42 J turns into useful work. In a nuclear plant, for every joule of heat produced, 0.3 J turns into useful work. In both cases the discarded energy warms up the surroundings.

Chapter

# 5

# Optics

## MICROSCOPE CONSTRAINTS

Refracting telescopes and microscopes have the same basic combination of lenses. There is an objective lens that produces a real image of the object, and an eyepiece lens that acts as a magnifying glass to examine the real image. Each lens is actually composed of several pieces of glass, designed to reduce spherical and chromatic aberration. There are geometrical constraints on the choice of focal lengths of the lenses. For the telescope, the object distance is large. The longer the focal length of the object lens, the larger the real image (although the image will still be much smaller than the object). However, the focal length of the object lens determines the length of the telescope. If the telescope is too long, it cannot conveniently be held. Therefore, the focal length of the objective of a hand-held telescope should be shorter than about 50 cm.

With a microscope, the geometrical constraints are just opposite to those of a telescope. The objective lens of a microscope can approach very close to the object, and the object distance can essentially be the focal length of the lens. The image distance is thus large and the real image is larger than the object. However, the microscope must be manipulated by a person sitting down and cannot conveniently be longer than about 25 cm. Therefore, the magnification of the object stage is

$$(\text{image distance})/(\text{object distance}) = 25/f_o.$$

Without a magnifying glass we could comfortably look at that real image from a distance of 25 cm from the eye. With a magnifying glass held at the eye, we can approach the object to a distance of $f_e$. Thus the eyepiece provides a magnification of 25 $/f_e$. The combined magnification is

$$\frac{(25)^2}{f_o f_e}.$$

To design a microscope with a magnification of 1000, choose an objective lens of 2.5 mm for a first-stage magnification of 100. The lens will be almost touching the specimen. The eyepiece can have a focal length of 2.5 cm, yielding a second stage magnification of 10.

Why not make an eyepiece with a focal length of 2.5 mm and get an overall magnification of 10,000? The field of view of such an eyepiece would be very small. More important, the wavelength of visible light is about half a micron. Resolution becomes impossible if we try to see things with a size comparable to or smaller than the wavelength of our probe. With a magnification of 1000, objects with a size of 1 micron appear to be 1 mm. That's about the limit of what the unaided eye can see. We can do better by about a factor of 2 by using blue light instead of red light ($4 \times 10^{-7}$ m versus $7 \times 10^{-7}$ m).

## BINOCULAR SIZE AND POWER

Binoculars and telescopes are frequently rated in terms of "power." That's magnification power, defined by the ratio of subtended angle of image to subtended angle of object. With the naked eye, the moon subtends an angle of one-half degree. With a 10-power telescope, it would subtend an angle of 5 degrees.

The equation for magnification of a telescope is $M = \frac{f_1}{f_2}$, where $f_1$ is the focal length of the objective lens, and $f_2$ is the focal length of the eyepiece. Evidently, for maximum magnification we want a long objective lens and a short focal length eyepiece. For distant objects, the obective lens forms an imagc at its focus, and the eyepiece acts like a magnifying glass with which you can examine that image. The shorter the focal length of a magnifying glass, the closer the eye can come to the image, and the larger the subtended angle.

The length of hand-held telescopes is limited to the length a person can hold. With eyepieces, on the other hand, there are optical and economic factors that mitigate against focal lengths of less than a couple of centimeters. Not only does the eyepiece have to have short focal length, it has to have a large diameter to provide a large field of view. To consider the engineering compromises, let's design a 10-power binocular. Choose an eyepiece with $f_2 = 2.5$ cm (which gives the eyepiece a power of about 10). Choose an objective lens with $f_1 = 25$ cm. The magnification of the objective lens by itself is just 1. The image it produces at its focus has the same angular width when viewed with the naked eye as does the object itself. (If the image of the moon were cast on a screen, the image, as seen from 25 cm away, would appear to be the same size as the moon.) However, because the image is now available for inspection by the eyepiece, with an eye-to-image distance of only 2.5 cm, the combined magnification of the arrangement is 10.

In the case of binoculars, a length of 27.5 cm (nearly a foot) would be awkward for the viewer. The optical path is partially shortened because the light travels part of the way inside the binoculars in glass prisms where the index of refraction is 1.5. The prisms also provide a folded path that shortens the outside length. Even with these tricks, it's impractical to design hand-held binoculars with a magnification much greater than 10 because it's hard to hold a 10-power binocular with a steady hand.

# SEEING UNDER WATER

An early lesson in learning to swim is to open one's eyes under water. It must be a little frightening for a child because one can't see very well under water, even with eyes open. The lens of a healthy eye is a remarkable organ, canceling out imperfections due to spherical and chromatic aberrations. But the major focusing actually takes place at the boundary between air and cornea. The lens just makes second-order corrections to focus the light on the retina. When you are seeing under water, there is very little refraction as the light leaves the water and passes through the cornea. The index of refraction of the cornea and the liquid in the eye is very close to that of water.

Optometrists describe the focusing power of lenses in terms of *diopters*, $D = \frac{1}{f}$, where $f$ is the focal length measured in meters. The smaller the focal length of the lens, the more "powerfully" it focuses. A converging lens has positive power in diopters, and a diverging lens has a negative $D$. One advantage of this system is that for thin lenses that are close together, the power in diopters just adds linearly:

$$\frac{1}{f} = \frac{1}{f_1} + \frac{1}{f_2}, \qquad D = D_1 + D_2.$$

The length of the human eye, from cornea to retina, is about 2 cm. Therefore, the optical power of the eye system is $D_{\text{eye}} = 1/0.02 = 50$. The major share of that power is provided by the curved surface of the cornea. It accounts for about $+35D$, and the lens fine-tunes the other $+15D$ for parallel light or to produce a focus for diverging light coming from closer objects.

If the lens provides $+15D$ to focus parallel light coming from infinity, it must provide more power to focus the diverging light coming from a near object. A normal eye can maintain a focus for objects up to about 30 cm. For this geometry of object ($o$), image ($i$), and focal distance ($f$),

$$\frac{1}{f} = \frac{1}{0} + \frac{1}{i} \Rightarrow \frac{1}{f} = \frac{1}{0.30} + \frac{1}{0.02} \Rightarrow D_{\text{system}} = 3.33 + 50.$$

The lens must increase its power from +15 to +18.3.

Consider what happens under water. Without the interface between air and cornea, there is very little refraction at the cornea. The focusing power of the eye then depends on the lens, with its maximum power of $D = +18.3$. That's a focal length of 0.055 m. Parallel light would be focused about 3.5 cm behind the retina. Everything under water would look blurred. You can cure this by wearing swimming goggles. The mask provides a plane surface between air and water, allowing normal seeing.

Is a near-sighted person better off under water? A person with 20/100 vision requires compensating glasses with $D$ about −6. Note that the minus sign means that the lens is diverging. The eye lens itself is too strong. In air, the power of the system with glasses is

$$D = -6 + 35 + D_{\text{eye lens}} = 50.$$

Therefore, the minimum power of the eye lens is +21D.

Under water, without glasses, the maximum power of this myoptic eye is +24.3. The focal length is about 4 cm. That's a little better than the situation with normal eyes, where the focal length is 5.5 cm. However, under water, things will still look fuzzy.

## CANDLE POWER

The motto of the Christophers is: "It is better to light one candle than to curse the darkness." A common candle gives off about the same amount of light as a flashlight bulb, but it gives off a lot more heat. Let's put some numbers in and see how much power we can get out of a candle.

During one hour, a candle with a diameter of 2 cm will reduce in height by about 2 cm. That's about 6 $cm^3$ of a hydrocarbon, or about 1 mole, that has been consumed.

The density of candle wax is about 1 $g/cm^3$. (A candle will just about float in water.) Therefore, the candle burns about 8 g/hr. The heat of combustion for a hydrocarbon (see the section "Molar Energy" in chapter 9) is about $50 \times 10^3$ J/g. In one hour, the burning candle will consume 8 g, producing $400 \times 10^3$ J. The power generated is $\frac{4000 \times 10^2 \text{ J}}{3600 \text{ s}} \approx 100$ W. That's about the same power radiated by a human at rest. (The original definition of a standard candle specified that it be made of whale oil and consume 120 grains = 7.8 g/hr.)

Chapter

# 6

# Electricity

## ELECTROSTATIC CHARGE ON A BALLOON

How much electric charge can you put on a blown-up balloon? Is the balloon at high voltage? Measure and find out! Blow up two spherical balloons, tie each with thread, and suspend the two from a common point. Choose a dry day to do this, and blow up the balloons in advance. The geometry is shown in figure 6.1.

Rub the balloons on your hair or a wool sweater. They will repel each other. A typical value of the angle of suspension from the vertical is 30°. The mass of a typical balloon is 4 g. The weight is $4 \times 10^{-2}$ N. The horizontal component is $2 \times 10^{-2}$ N, and the typical distance of the separation of centers is 40 cm. The repulsion between the balloons is provided by the Coulomb force:

$$0.02 = k\frac{q^2}{d^2} = 9 \times 10^9 \, \frac{q^2}{0.4^2}$$

$$q = 5 \times 10^{-7}\ \text{C} \approx \frac{1}{2}\,\mu\,\text{C}.$$

If we rub the balloon in such a way that the static charge is distributed fairly uniformly over the surface, we may use Coulomb's law for point or spherical sources.

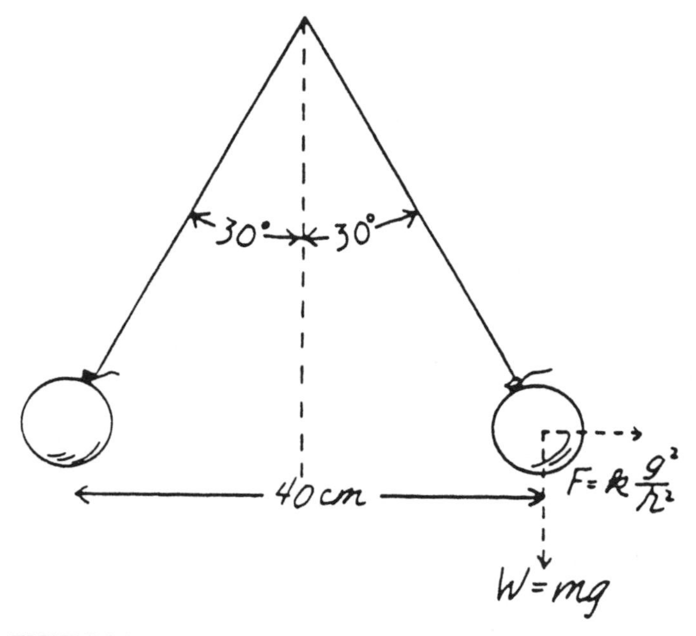


FIGURE 6.1 Geometry of suspended balloons, repelling each other because of their static charges.

The capacitance in pF of a sphere is approximately equal to the radius in centimeters. For our typical balloon, the radius is 10 cm and the capacitance is thus equal to $10^{-11}$ F. With a charge of $10^{-7}$ C, the electrostatic potential of the balloon is

$$V = \frac{q}{C} = \frac{5 \times 10^{-7}}{10^{-11}} = 50{,}000 \text{ V !}$$

## CAPACITORS

Most capacitors in electrical circuits consist of parallel plates:

$$V_{\text{across plates}} = 4\pi k \frac{Q}{A} d = \frac{Qd}{\varepsilon A} = \frac{Qd}{\kappa \varepsilon_0 A}.$$

The spacing between the plates is $d$. The area of the plates is $A$. The permittivity of free space is $\varepsilon_0 = \frac{1}{4\pi k} = 8.85 \times 10^{-12}$ (farad meter). The dielectric constant for a particular material is $\kappa$. The capacitance of a parallel plate capacitor is thus

$$C = \frac{Q}{V} = \frac{\kappa \varepsilon_0 A}{d}.$$

Practical capacitors exploit one or another of these parameters. A paper and foil capacitor consists of a spool of two layers of foil and one very thin sheet of insulator sandwiched together and rolled up. The total surface area is large, and the spacing is just the thickness of the thin insulating paper or plastic. In electrolytic capacitors, the surface area of the foil is not very large, but the spacing between the conducting foil is only a few molecules thick, provided by a chemical barrier with a conducting paste that serves as the other plate. In these, the separation materials have a dielectric constant only a few times unity. In a ceramic capacitor, the area is small, the spacing is not very small, but the dielectric constant of the material can be as high as 10,000.

Let's calculate the capacitance of a ceramic capacitor that has an area of about 1 $cm^2$ and a spacing of a millimeter:

$$C = \frac{(10^4)(8.85 \times 10^{-12})(10^{-4})}{(10^{-3})} \approx 10^{-8}\,\text{F} = 0.01\,\mu\text{F}.$$

Some low-voltage capacitors on the market have extremely high capacitance. In a few cubic centimeters they present a capacitance of one or several farads. They act almost like batteries. However, they are true capacitors, as can be demonstrated by observing their RC time constants (greater than one second) and decay behavior. They must have an interior structure much like that of "activated carbon," where connecting honeycomb cavities provide a huge surface area. The spacing must be like that of an electrolytic capacitor in which a molecular layer is deposited between the solid electrode and a paste for the other electrode. Let's assume a spacing of the width of 10 molecules and calculate the area that must be present. We will also assume that $\kappa = 1$.

$$C = 1\ \mathrm{F} = \frac{(1)(8.85 \times 10^{-12})A}{(10 \times 10^{-9})} \Rightarrow A \approx 10^{3}\ \mathrm{m}^{2}.$$

That's an area of about a quarter acre!

## CAPACITANCE OF SPHERES

The electrical capacitance of an object is the ratio of its electric charge to its potential: $C = \frac{Q}{V}$. There is a particularly simple relationship for the capacitance of a sphere, since $V = k\frac{Q}{R}$. Therefore, $C_{\text{sphere}} = \frac{R}{k}$. This can be transformed numerically to become

$$C = \frac{(R \text{ in m})}{9 \times 10^{9}} \approx 1 \times 10^{-10}\ (R \text{ in m})$$
$$= 1 \times 10^{-12}\ (R \text{ in cm}) = (R \text{ in cm})\ \mathrm{pF}.$$

Thus the capacitance of a sphere in picofarads ($10^{-12}$ F) is about equal to its radius in centimeters. A table-top Van de Graaff generator may have a bulb that is 10 cm in radius. Its capacitance is therefore about 10 pF. At a potential of 100,000 volts, it has a charge of $10^{-6}$ coulombs.

The capacitance of the Earth is surprisingly small:

$$C = (6.4 \times 10^{8}\ \mathrm{cm})\ \mathrm{pF} = 640\ \mu\mathrm{F}.$$

This rule-of-thumb formula provides good approximations for the capacitance of small objects even if they are not spheres. A 10 cm circuit wire in a metal chassis has a capacitance of about 10 pF. When you shuffle over a new carpet or throw back a woolen blanket in the wintertime, you can draw sparks several centimeters long. In dry weather, the potential difference that produces sparks is about 12,000 volts per centimeter. If you generate a one-inch spark, your potential is about 30,000 volts. You might model yourself as a 60 cm sphere. Your capacitance is about 60 pF, and the charge you had before the spark was $2 \times 10^{-6}$ C.

The electric potential and the charge on a balloon are calculated in the section "Electrostatic Charge on a Balloon" at the beginning of this chapter.

## ELECTRIC FIELD IN A WIRE

Before an electric circuit is closed, the wires are charged positive or negative, and the electric field lines are radial toward or away from the cylindrical wires. As soon as the circuit is completed, the field lines lie in the metal, following the wire through twists, turns, or knots. What guides the field to turn a corner such as the one shown in the diagram? There must be positive and negative surface charges on the wire (fig. 6.2).

Take a typical battery circuit with 1 A in copper wire of 1 $mm^2$ cross section. A handbook value for conductivity shows that to main-

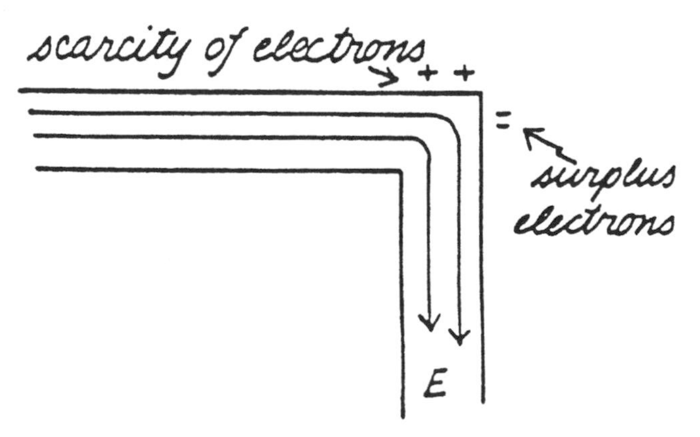


FIGURE 6.2 Electric field lines in a current carrying wire.

tain this current requires 17 mV across 1 m. Therefore, the field in the wire is

$$E = \frac{17 \times 10^{-3}\ \text{V}}{1\ \text{m}} = 17 \times 10^{-3}\ \text{V/m}.$$

Where the field lines run into the right-angle turn, there must be a surface charge:

$$\sigma = \varepsilon_0 E = (8.9 \times 10^{-12})(17 \times 10^{-3}) = 1.5 \times 10^{-13}\ \text{C/m}^2.$$

The total charge toward which the $E$ lines are aimed is

$$Q = \sigma A = (1.5 \times 10^{-13}\ \text{C/m}^2)(1 \times 10^{-6}\ \text{m}^2) = 1.5 \times 10^{-19}\ \text{C}.$$

By the fortuitous choice of parameters, we have arrived at a value for $Q$ equal to one electron charge. Interestingly, a wire carrying such a current needs only about one electron, in excess or deficit, near right-angle corners, in order to produce the electric field necessary to maintain the current. In actual fact, these surface charges are only quasi-static, subject to continual fluctuations.

## ELECTRON DRIFT SPEED

The Bohr atom is a surprisingly useful model for describing the behavior of the electron in a hydrogen (or alkali) atom. A similar situation exists in the models of electron movement in electric circuits. At the beginning of the twentieth century, Paul Drude worked out a similarly effective theory based on the idea that electrons in a metal behave like molecules in a gas. Each atom contributes one or two electrons to the system. The electrons would share thermal kinetic energy of $\frac{3}{2}kT$ and would dash about, running into the stationary ions of the metallic lattice. If an electric field were imposed, the electrons would accelerate until they next collided with an ion. These interrupted accelerations would produce a drift velocity proportional to the strength of the field. Let's calculate that drift velocity for electrons in a flashlight circuit.

Figure 6.3 shows the geometry of our calculation. All the electrons in a cylinder of length $v_{\text{drift}}\Delta t$ and cross-sectional area $A$ will pass the measuring point in the time $\Delta t$. The density of free electrons in the metal is $n$ and the charge density is $ne$. The amount of charge in the cylinder is thus $\Delta q = neAv_{\text{drift}}\Delta t$. The current is $i = \frac{\Delta q}{\Delta t} = neAv_{\text{drift}}$. To find the drift velocity, we must calculate the free electron density. Copper provides one conduction electron per atom. The number of electrons per $\text{cm}^3$ is

$$n = \left(\frac{1 \text{ charge carrier}}{\text{atom}}\right)\left(\frac{\text{atoms}}{\text{mole}}\right)\left(\frac{\text{mole}}{\text{gram}}\right)\left(\frac{\text{gram}}{\text{cm}^3}\right)$$

$$= (1)(6 \times 10^{23})\left(\frac{1}{64}\right)(9) = 8.4 \times 10^{22} \text{ charge carriers/cm}^3.$$

This number is the same as the density of atoms in a solid, about 2000 times greater than the density of gas molecules at room temper-

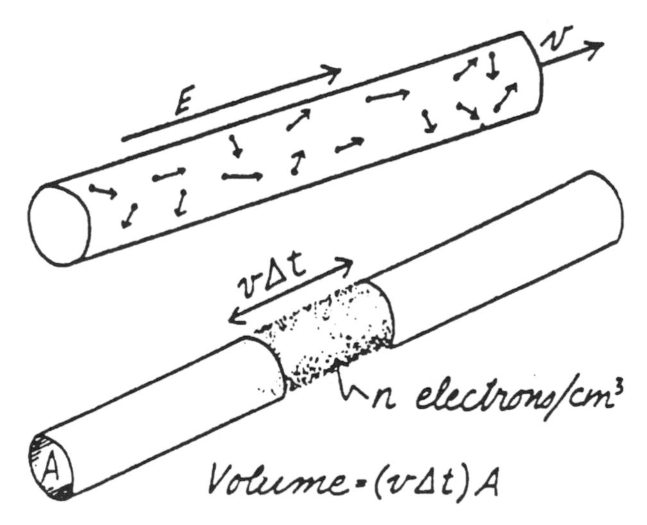


FIGURE 6.3 Geometry for derivation of drift velocity of electrons in current carrying wire.

ature. Surely the model of an electron gas must break down, particularly because of the strong long-range interactions between electrons. Nevertheless, let's calculate the drift velocity for a current of 1 A in a copper wire with a cross section of 1 $\text{mm}^2$:

$$v_{\text{drift}} = \frac{1\ \text{A}}{(8.4 \times 10^{22}\ \text{carriers/cm}^3)(1.6 \times 10^{-19}\ \text{C/carrier})(1 \times 10^{-2}\ \text{cm}^2)}$$
$$= 7 \times 10^{-3}\ \text{cm/s}.$$

The *rms thermal* velocity of electrons in such a gas at room temperature is

$$v = \sqrt{\frac{3kT}{m}} = \sqrt{\frac{3(1.38 \times 10^{-23})(293)}{(9.1 \times 10^{-31})}} = 1.2 \times 10^5\ \text{m/s}.$$

Evidently the electron drift velocity is just a small perturbation imposed on the basic random motion. The actual ratio of velocities is even greater than we calculate with the Drude model. Because of quantum requirements, the electrons do not share the thermal energy distribution of the ions of the lattice. Instead, they are organized in closely spaced energy levels up to the Fermi-level energy. The speed corresponding to a Fermi energy of several eV is about $2 \times 10^6$ m/s. Nevertheless, the drift velocity calculated with this crude model agrees with experiments such as the Hall effect. When a circuit is first closed, the electric field propagates in the wire with a speed close to that of light. That field starts the drift velocity of electrons all along the circuit.

## UNFAMILIAR CURRENTS

We usually know the voltage across everyday devices, but we seldom consider the currents involved. For instance, a 60-watt bulb in a lamp has 120 volts across it. To be sure, the voltage is usually alternating and the 120 is smaller by a factor of $\sqrt{2}$ than the peak amplitude of 170. The current is also alternating, but its root-mean-square value is

$$I = \frac{power}{V} = \frac{60 \text{ W}}{120 \text{ V}} = \frac{1}{2} \text{ A}.$$

The resistance of the filament must be

$$R = \frac{V}{I} = \frac{120 \text{ V}}{\frac{1}{2} \text{ A}} = 240 \ \Omega.$$

If you check this out with a battery driven ohm-meter, you will find that the filament resistance is only 20 Ω. The 1.5 or 3 V battery in the ohm meter does not produce enough energy to raise the filament temperature appreciably. However, as soon as the 120 V is put across the filament, there is a current surge starting at 6 A that rapidly raises the temperature and the resistance.

The temperature dependence of the resistance of a tungsten wire is given by

$$R = R_o(1 + \alpha t) = R_o + R_o(0.0045)t.$$

$R_o$ is the resistance at room temperature, and $t$ is the number of Celsius degrees above room temperature:

$$240 = 20 + 0.090t \quad \Rightarrow \quad t = 2440°\text{C}.$$

The operating temperature of the filament is therefore about 2700 K.

Many flashlight bulbs are rated at 3 W. With two 1.5 V cells in series, the current is $I = \frac{3 \text{ W}}{3 \text{ V}} = 1$ A. Notice that this is twice the current in a 60 W bulb.

A kitchen toaster is often rated at 1 kW. With the standard 120 V, the current is 8.3 A. Since many home circuits are fused for 15 A, don't use a toaster and a 1 kW hair dryer on the same circuit.

A car battery has thick copper leads going to the starter motor. That's because a medium-size car has a starting motor requiring one horsepower, which is 746 W. With 12 V furnished by the battery, the current is about 60 A, and thick leads are necessary so that not much voltage is lost between battery and motor.

## HIGH COST OF BATTERY ELECTRICITY

Compared with a dry cell, electricity from the wall is dirt cheap. Consider, for example, the cost and energy capacity of a D cell. The battery costs a dollar, and can provide $^2/_3$ A at 1.5 V for about 5 hours. Thus, the energy provided is 5 watt-hours. If the battery costs one dollar, the cost of a watt-hour is 20 cents, and the cost of a kilowatt-hour would be $200.

A kilowattt-hour from the wall costs only about fifteen cents. Of course, there are occasions when you would rather pay more and not use a long extension cord. (A similar calculation is made in Robert Romer's text, *Energy, an Introduction to Physics*, p. 299 W. H. Freeman, 1976.)

## MAGNET STRENGTH

A ferromagnet is composed of individual atomic magnets that are all lined up. (To be sure, the atoms in each domain are lined up with each other, and the domains are mostly lined up.) If we can calculate the strength of an individual atomic magnet, we ought to be able to calculate the strength of a full-scale magnet.

Assume a Bohr model of an atomic current loop, with magnetic moment $\mathbf{m} = IA$. The current is provided by a circulating electron with rotational period, $T$.

$$I = \frac{e}{T} = \frac{ev}{2\pi r}$$

$$\mathbf{m} = \frac{ev}{2\pi r}(\pi r^2) = \frac{evr}{2} = \frac{e}{2m}\,(mvr).$$

But the angular momentum of the circulating electron $= L = mvr$. Therefore, $\mathbf{m} = \frac{e}{2m}L$.

Now we bypass our Bohr model and note that $L$ must be quantized, with units of Planck's constant, $\hbar$. The smallest value for $m$ is

$$\mathbf{m} = \frac{e}{2m}\hbar$$

$$= \frac{(1.6 \times 10^{-19}\ \mathrm{C})}{2\,(9 \times 10^{-31}\ \mathrm{kg})}\,(1 \times 10^{-34}\ \mathrm{kg} \cdot \mathrm{m}^2\mathrm{s}^{-1}) \approx 1 \times 10^{-23}\ \mathrm{A} \cdot \mathrm{m}^2.$$

For a mole of such atoms all lined up, $\mathbf{M} = (6 \times 10^{23})(1 \times 10^{-23}) = 6\ \mathrm{A} \cdot \mathrm{m}^2$.

Indeed, that's about the magnetic moment of a good bar magnet with a volume of about 10 $\mathrm{cm}^3$ (containing a mole of atoms). Of course, it ignores the fact that the atomic magnetic moments are created by the spins (not rotations) of shielded electrons in the atoms. In an iron atom, four interior electrons with parallel spins contribute to the magnetic moment. Not all of the individual domains line up, however.

This explanation of magnetic strength faces a serious objection, however. Why doesn't thermal energy prevent atomic alignment? The dipole energy is $\Delta U = -2\mathbf{m} \cdot \mathbf{B} = -2(1 \times 10^{-23})(1\ \text{tesla}) = -2 \times 10^{-23}\ \mathrm{J} \approx 1 \times 10^{-4}\ \mathrm{eV}$. Room temperature thermal energy of an atom is about $^1/_{25}$ eV, much more than is necessary to prevent alignment of the magnetic dipoles. Evidently, the alignment is maintained by some agency other than the mutually generated magnetic field. It's a quantum effect of the electron systems of neighboring atoms, adjusting their orientation to minimize energy while satisfying the Pauli exclusion principle.

## MATCHING EARTH'S B FIELD

Besides the Earth's fairly uniform gravitational field, there is a magnetic field that has about the same magnitude everywhere on the surface. To be sure, the *direction* of the field is not uniform. At the equator, the field is tangential to the surface, and at the (magnetic) poles the direction is vertical. The magnetic field strength at the equator is about 0.3 gauss, or $0.3 \times 10^{-4}$ tesla. At the magnetic poles it is about 0.7 G. To a first approximation the field acts as if there were a giant dipole in the Earth, one that is not quite aligned with the Earth's axis and whose orientation drifts by several degrees per century.

At a latitude of 40° the horizontal component of the magnetic field is about 0.23 G. We can match this weak field with a dry cell and a small length of wire. If the lab field is produced perpendicular to the Earth's horizontal field, the angle of the resultant field can be measured with a small magnetic compass.

First, send about 5 A through a long straight wire. The source of this current can be a D cell short-circuited by the wire. For a few seconds it won't ruin the battery. The field produced at a distance $r = 2$ cm from a long straight wire is

$$B = \frac{\mu_0 i}{2\pi r} = \frac{(4\pi \times 10^{-7})\ 5\ \text{A}}{2\pi(0.02\text{m})} = 5 \times 10^{-5}\ \text{T} = 0.5\ \text{G}.$$

By increasing the distance to 4 cm or reducing the current to 2.5 A, you can arrange to produce a magnetic field equal to the Earth's horizontal field. The compass needle will swing to 45°. This demonstration duplicates the classical discovery by Oersted that magnetism and electricity are related. Oersted noticed the effect during a lecture when a current-carrying wire lay on top of a compass.

Alternatively, you can make a single turn coil with the wire, looping it around the compass so that its axis is horizontal and perpendicular to the Earth's horizontal field. The field at the center of this loop is

$$B = \frac{\mu_0 i}{2r} = \frac{(4\pi \times 10^{-7})\ 5\text{A}}{2(0.04\text{m})} = 8 \times 10^{-5}\ \text{T} = 0.8\ \text{G}.$$

Once again, you can vary the parameters, particularly the current, to produce a field equal to that of the Earth's horizontal field, thus making the compass needle turn to 45°.

The Earth's field is produced in some (still unknown) way by rotation of the liquid iron core. The field varies over the centuries, not only in direction but also in strength. It even changes direction every so often! The magnetic dipole moment of a model that would produce our current field is given in most handbooks in c.g.s. units: $8 \times 10^{25}$ gauss cm$^3$ = $8 \times 10^{15}$ T m$^3$. In SI units this is equal to $8 \times 10^{22}$ A m$^2$. The liquid (and conducting) core extends out to a radius of $3 \times 10^6$ m. The current produced by the rotating core must be

$$I = \frac{\mathbf{M}}{A} = \frac{(8 \times 10^{22}\ \mathrm{A \cdot m^2})}{\pi(3 \times 10^6\ \mathrm{m})^2} = 3 \times 10^9\ \mathrm{A}.$$

That seems like an enormous current. Consider, however, that on the axis of such a ring current, at a distance $r$ from the plane, the magnetic field would be

$$B = \frac{2\mu_o}{4\pi}\frac{\mathbf{M}}{r^3} = (2 \times 10^{-7})\frac{8 \times 10^{22}}{(6 \times 10^6)^3} = 7.4 \times 10^{-5}\ \mathrm{T} = 0.7\ \mathrm{G}.$$

This formula for B on the perpendicular axis of a dipole is strictly good only when $r >> R$, the radius of the current loop. Note, however, that the crude model gives surprising agreement with the value for the field at the pole.

## HANGING WIRE

Parallel electrical currents attract each other. Can you stretch a wire underneath another wire so that the one underneath is supported against its weight and sticks to the upper one? The force per meter between two parallel wires that touch each other is

$$\frac{F}{L} = \frac{\mu_o}{2\pi}\frac{i^2}{2r}.$$

The weight of the suspended meter of wire is $\pi r^2 \rho g$.

$$\frac{\mu_o}{2\pi}\frac{i^2}{2r} = \pi r^2 \rho g \quad \Rightarrow \quad i = 2\pi\sqrt{\frac{\rho g}{\mu_o}}\, r^{\frac{3}{2}}$$

$$= 2\pi\sqrt{\frac{(8.9 \times 10^3 \text{ kg/m}^3)(9.8 \text{ N/kg})}{4\pi \times 10^{-7}}}\, r^{\frac{3}{2}}$$

$$i = (1.66 \times 10^6)\, r^{\frac{3}{2}}.$$

We have used the density of copper. For wires with radius of 1 mm, the current in each wire is 52 A. Note the dependence of the current on the wire radius. If the wire had a radius of 2 mm, the required current would be 147 A, clearly prohibitive for casual demonstration. A current of 52 A, however, can be produced, with caution, from a car battery. The use of aluminum wire would reduce the required current by $\sqrt{(2.7)/(8.9)} = 0.55$, to a more comfortable 29 A.

The resistance of a one-meter copper wire with a radius of 1 mm is

$$R = \rho\frac{L}{A} = (1.7 \times 10^{-8}\ \Omega \cdot \text{m})\frac{1 \text{ m}}{\pi(1 \times 10^{-3})^2} = 5.4 \times 10^{-3}\ \Omega.$$

The power dissipated in the wire carrying a current of 52 A is

$$P = (52 \text{ A})^2\, (5.4 \times 10^{-3}\ \Omega) = 14.6 \text{ W}.$$

The wire would get hot, but it would not melt.

## ENERGY STORAGE IN L AND C

One of the functions of inductances and capacitors in circuits is the momentary storage of energy. The energy stored in each is

$$U_{\text{C}} = \frac{1}{2}CV^2, \quad U_{\text{L}} = \frac{1}{2}LI^2.$$

A high-voltage 1 $\mu F$ capacitor charged to 2000 volts has a stored energy of

$$U = \frac{1}{2}(1 \times 10^{-6}\ \text{F})(2 \times 10^{3}\ \text{V})^{2} = 2\ \text{J}.$$

That doesn't seem like much energy. An oil-filled 1 $\mu F$ capacitor that can stand 2000 volts has a mass of about 500 g. If it were raised only 40 cm in the Earth's field, it would have as much gravitational as electrical energy. However, with the right circuitry (perhaps achieved by putting your hand across the terminals) you could extract the electrical energy in a microsecond. The momentary power would be $2 \times 10^{6}$ W.

A common laboratory air-core solenoid has an inductance of 1 henry and a resistance of 50 ohms. If we use a 6 V battery, the equilibrium current is $I = 0.12$ A. The stored magnetic energy is

$$U = \frac{1}{2}(1\ \text{H})(0.12\ \text{A})^{2} = 0.0072\ \text{J}.$$

The solenoid has a mass of about 2 kg. The stored magnetic energy would hardly be enough to lift the solenoid 0.4 mm.

Since the inductive energy depends on the square of the current, let's consider a magnet that uses a lot of current. First, let's make a detour to derive a more useful formula for stored magnetic energy. In a solenoid the energy density is equal to the stored energy divided by the volume of the solenoid:

$$u_{\text{B}} = \frac{U_{\text{B}}}{V} = \frac{\frac{1}{2}LI^{2}}{V} = \frac{\frac{1}{2}\mu_{\text{o}}n^{2}VI^{2}}{V} = \frac{1}{2}\mu_{\text{o}}n^{2}I^{2}.$$

The inductance of a long solenoid is $L = \mu_{o}n^{2}V$, where $n$ is the number of turns per meter. The magnetic field in a solenoid is $B = \mu_{\text{o}}nI$. Now we can express the energy density in terms of the magnetic field:

$$u_{\text{B}} = \frac{1}{2}\frac{B^{2}}{\mu_{\text{o}}}\ \text{J/m}^{3}.$$

A typical research magnet used for beam separation in high-energy accelerators produces a magnetic field of 1.5 tesla. The energy is stored mostly in the magnet gap, since the energy density in the iron is smaller by a factor of $\mu_o/\mu$. For iron this permeability ratio is about 1/1000. If the volume of the gap is about $^1/_5$ m$^3$, the stored energy is

$$U_B = \frac{1}{2}\frac{B^2 V}{\mu_o} = \frac{1}{2}\frac{(1.5\ \mathrm{T})^2(0.2\ \mathrm{m}^2)}{(4\pi \times 10^{-7})} = 1.8 \times 10^5\ \mathrm{J}.$$

That's the energy in only a few cubic centimeters of gasoline. However, if you accidentally open the circuit carrying the current, that magnetic energy will form an enormous spark, melting everything in its path.

## JUMP-ROPE GENERATOR

You can generate electricity by twirling a wire in the Earth's magnetic field. Choose a light wire about 8 m long, plugging the two ends into a millivoltmeter or oscilloscope that is sensitive in the millivolt region. Swing the middle 4 m of the wire like a skipping rope, creating a rotating loop with a diameter of about 3 m. With a period of 1 s, the induced emf should be

$$E = -\frac{d\Phi}{dt} = -B\frac{dA}{dt} = (0.4 \times 10^{-4}\ \mathrm{T})\frac{\pi(1.5\ \mathrm{m})^2}{1\ \mathrm{s}} = 0.3\ \mathrm{mV}.$$

Such a low voltage is detectable, but several factors can be improved to increase the emf: use a longer wire; spin the wire faster; orient the long axis of the loop in the east-west direction. This will increase the maximum flux through the loop, although the vertical component of the field will provide the major flux regardless of the horizontal orientation. If you use a millivoltmeter, the needle may not be able to respond to the rapid voltage changes. If you use an oscilloscope that has a millivolt range, you should terminate the input with 100 Ω or so to reduce the pickup of electrical noise.

## Q OF A CRYSTAL RADIO

Building a crystal model radio is something that Cub Scouts do with a kit from Radio Shack. To design and build one from scratch, however, is not child's play. Consider the problem. The electromagnetic field from a radio station 20 miles away is several mV/m. The carrier signal, with a frequency of about 1 MHz, must be rectified so that the amplitude modulations can affect the headset. But all rectifiers at these frequencies, whether they are solid state or tubes, require signals in the fractional volt range. The transition between forward conduction and backward high resistance occurs as shown in figure 6.4. Somehow, the mV signal must be amplified by a factor of at least 100.

There is another constraint. The bandwidth for AM radio is 10 KHz. This narrow range provides for modulation of $\pm$ 5000 Hz, which is enough for low-fidelity sound. The narrow bandwidth allows many stations to operate in the same region, but the receiver must be able to respond to the carrier frequency of a station without overlap from another station. The frequency response must look like the one in figure 6.5. This frequency response is characterized by the Q or "quality factor" of a resonant circuit:

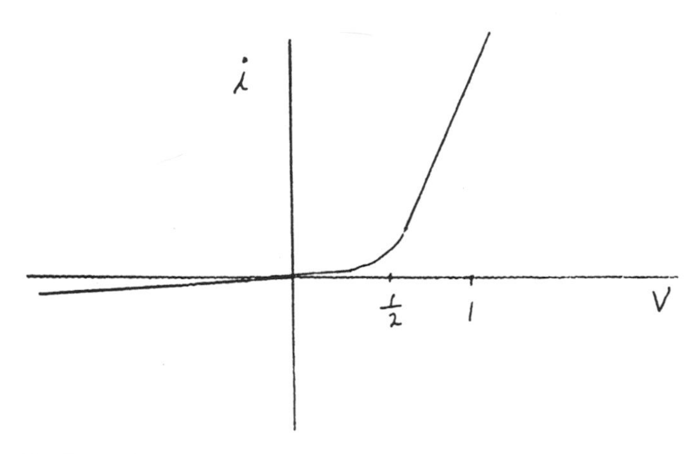


FIGURE 6.4 Current as a function of voltage for a diode.

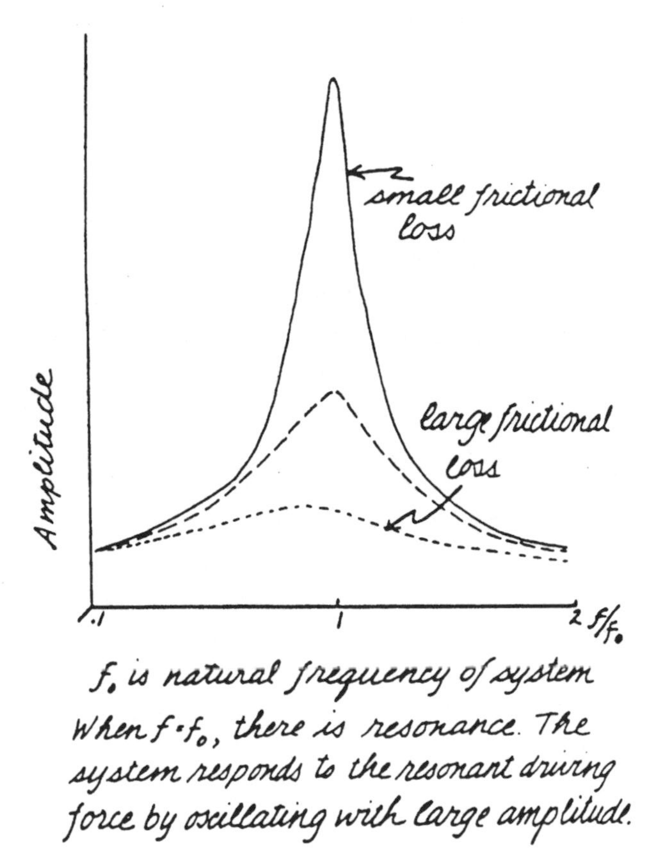


FIGURE 6.5 Amplitude as function of frequency for three different resonance systems.

$$Q = \frac{f_0}{\Delta f}, \text{ where } f_0 \text{ is the resonant frequency}$$

and $\Delta f$ is the width of the resonance.

For AM radio, $f_o = 1 \times 10^6$ Hz in the center of the band spread between 535 and 1605 Hz. The $\Delta f$ is chosen to equal 10,000 Hz. Therefore, for AM radio, $Q = \frac{10^6}{10^4} = 100$.

The Q of a circuit also gives the amplification of the voltage across the circuit. With a Q of 100, a 5 mV signal is amplified enough to be rectified. For a crystal radio with no extra amplification, the coincidence of the dual requirements of Q = 100 makes the system workable.

## E AND B FROM A LIGHTBULB

We are continually being bombarded with electromagnetic radiation! Is it safe? Are we in danger? Consider that ordinary lightbulbs emit light and therefore emit electric and magnetic fields. Let's put in the numbers.

The total energy density in an electromagnetic wave is the sum of the energy densities in the two fields:

$$\mu_{\text{total}} = \mu_{\text{E}} + \mu_{\text{B}} = \frac{1}{2}\varepsilon_o E^2 + \frac{1}{2}\frac{1}{\mu_o}B^2.$$

Since $c = \sqrt{\frac{1}{\varepsilon_o \mu_o}}$, and $B = \text{E}/\text{c}$,

$$\frac{1}{2}\frac{1}{\mu_o}B^2 = \frac{1}{2}\frac{1}{\mu_o}\frac{E^2}{c^2} = \frac{\varepsilon_o \mu_o}{2\mu_o}E^2 = \frac{1}{2}\varepsilon_o E^2.$$

The energy is evenly divided between electric and magnetic fields. The field intensity, $I$, has the units of watts per square meter ($\text{W/m}^2$). For a plane wave, $I = \varepsilon_o E^2 c$.

Let's calculate the strength of $E$ and $B$ at a distance of 3 m from a 100 W light bulb. The light-producing efficiency is only 2.5%. We will assume that 2.5 W are radiated uniformly in all directions. At a distance of 3 m, the radiation passes through an area of $4\pi\ r^2 = 4\pi(3\text{ m})^2 = 113\text{ m}^2$. Thus the watts per square meter at this distance is 0.022. Half of this is provided by the electric field and half by the magnetic field:

$$\frac{1}{2}\varepsilon_o E^2 c = \frac{1}{2}I = \frac{1}{2}(0.022\text{ W/m}^2)$$

$$E = \sqrt{\frac{0.022}{(8.85 \times 10^{-12})(3 \times 10^{8})}} = 2.9 \text{ V/m}.$$

That's a strong field compared with that of most radio and TV signals, which are only a few microvolts to millivolts per meter.

The magnetic field at this distance from the bulb is

$$B = \frac{E}{c} = \frac{(2.9 \text{ V/m})}{3 \times 10^{8}} \approx 1 \times 10^{-8} \text{ T} = 1 \times 10^{-4} \text{ gauss}.$$

Although the power in the magnetic field is equal to the power in the electric field, the magnetic field strength is evidently very weak.

The intensity of solar radiation at the Earth's surface is about 1 kW/m$^2$, which is larger than the radiation 3 m from the 100 W bulb by a factor of $4.5 \times 10^4$. The electric field is therefore larger by the square root of $4.5 \times 10^4$, which is about a factor of 200. From the Sun, $E = 600$ V/m, and $B = 0.02$ G.

## TOASTER POWER

A toaster oven has three rod elements, each 0.25 m long with a diameter of 4 mm. These are connected directly across the 120 (AC) house volts, and each is rated at 400 W. The current, evidently, is

$$I = \frac{400 \text{ W}}{120 \text{ V}} = 3.3 \text{ A}.$$

We explain the heating effect in terms of electrons colliding with the ions in the lattice of the conductor.

Another explanation of the power generated is in terms of the Poynting vector. The cross-product of an electric field and magnetic field gives the value of the electromagnetic power crossing each square meter of the crossed fields:

$$\vec{S}(\mathrm{W/m^2}) = \frac{\vec{E} \times \vec{B}}{\mu_0}.$$

The electric field along the heating rod is $E = \frac{120\ \mathrm{V}}{0.25\ \mathrm{m}} = 480\ \mathrm{V/m}$. The magnetic field encircling the rod at the periphery of the rod is

$$B = \frac{\mu_0}{2\pi}\frac{I}{r} = \frac{\mu_0}{2\pi}\frac{3.3\ \mathrm{A}}{2 \times 10^{-3}\ \mathrm{m}} = \frac{\mu_0}{\pi}\,825\ \mathrm{T}.$$

The surface area of the rod is $A = (0.25\ \mathrm{m})\ 2\pi(2 \times 10^{-3}\ \mathrm{m}) = \pi \times 10^{-3}\ \mathrm{m}^2$. The power flooding into the rod (perpendicular to $\vec{E}$ and $\vec{B}$) is

$$\frac{(\vec{E} \times \vec{B})}{\mu_0}A = \frac{(480\ \mathrm{V/m})}{\mu_0} \times \left(\frac{\mu_0}{\pi}825\ \mathrm{T}\right)(\pi \times 10^{-3}\ \mathrm{m}^2) = 400\ \mathrm{W}.$$

One might feel uneasy about the model because static electric and magnetic fields do not generate power, no matter how they are arranged. However, the electric field in a conductor is not static but is continually being rearranged by the current to provide a constant flow of charge. The flow of power called for by the Poynting vector persists only so long as the electromotive source maintains the potential difference across the rod.

## MAGNETIC RESONANCE IMAGING

This remarkable diagnostic tool used to be called *nuclear magnetic imaging*, but the word "nuclear" worried people. The process depends on flipping the spins of protons in the nuclei of hydrogen, and then listening to the radiation as the protons flip back to their lower energy position. To begin, the protons in the patient are partially aligned in a magnetic field of about 1 tesla (10,000 gauss). The protons have magnetic moments and so would completely line up in the field like a bunch of compass needles except that thermal excitation keeps the distribution with and against the field about equal. The energy of alignment is

$$U = 2\mathbf{m} \cdot \mathbf{B},$$

where **m** is the magnetic moment of the proton, and **B** is the strength of the magnetic field in tesla.

If a pulse of electromagnetic radiation strikes the protons, with a photon energy equal to $U$, some of the protons in the lower energy state aligned with $B$ will flip to the higher energy state opposed to $B$. Then the activated protons will settle back to the lower energy, emitting the resonant energy, $U$.

Let's calculate the frequency of this radiation. The magnetic moment of a proton is $1.4 \times 10^{-26}\ \mathrm{A \cdot m^2} = 1.4 \times 10^{-26}\,\mathrm{J/T} = 8.8 \times 10^{-8}\,\mathrm{eV/T}$.

$$U = 2\ (8.8 \times 10^{-8}\ \mathrm{eV/T})(1\ \mathrm{T}) = 1.8 \times 10^{-7}\ \mathrm{eV}.$$

The frequency associated with a photon having this much energy is

$$(1.8 \times 10^{-7}\ \mathrm{eV})(1.6 \times 10^{-19}\ \mathrm{J/eV}) = (6.6 \times 10^{-34}\ \mathrm{J \cdot s})\nu$$

$$\nu = 43\ \mathrm{MHz}, \quad \lambda = \frac{3 \times 10^{8}}{43 \times 10^{6}} = 7.0\ \mathrm{m}.$$

This frequency is just below the TV channels. In an MRI instrument, the patient lies on the axis of a solenoid that provides a main axial field of about one tesla. There is a large antenna underneath the patient that both radiates and detects the E-M signal. In spite of the long wavelength, the magnetic resonance images have precision in the millimeter range. This is accomplished by having auxiliary coils produce small axial magnetic fields that add to or subtact from the main field. There is a gradient in this extra field from plus to minus. In the zero cross-over region, the resulting field is just right for the resonant flipping. The auxiliary coils are activated to determine a cross-section slice. When the E-M field is pulsed, some of the protons in the slice are flipped. As they return to their lower energy state, they radiate the resonant signal that is picked up by the antenna coil. By activating other auxiliary coils, the resonant region can be moved in the plane perpendicular to the axis, thus locating a “pixel” with a precision of about a millimeter.

Chapter

# 7

# Earth

## RADIUS OF EARTH

Eratosthenes measured the radius of the Earth in Alexandria, Egypt, 2200 years ago. He knew that at noon on June 21 the sun shone directly down a deep well in southern Egypt where the Aswan Dam now stands. This area is on the Tropic of Cancer. At the same time, a vertical rod in Alexandria cast a shadow about equal to one-twelfth the length of the rod. The geometry is shown in figure 7.1. The distance between Alexandria and the well was known very precisely because the Egyptians had government pacers to map the land each year after the Nile flooded. In our units, the distance was about 500 miles. Therefore, a quadrant of the Earth's circumference must be 6000 miles, and the whole circumference is 24,000 miles, in remarkable agreement with our modern measurements.

If you know some geography, there is a cruder way of figuring out the size of the Earth. The distance between New York and San Francisco is about 3000 miles. This number must be about right because it takes a jet five to six hours to make the trip, depending on head winds. Since a passenger jet cruises at 500 to 600 miles per hour, the numbers agree. As we know, there is a three-hour time zone difference between coasts. That's about one thousand miles per

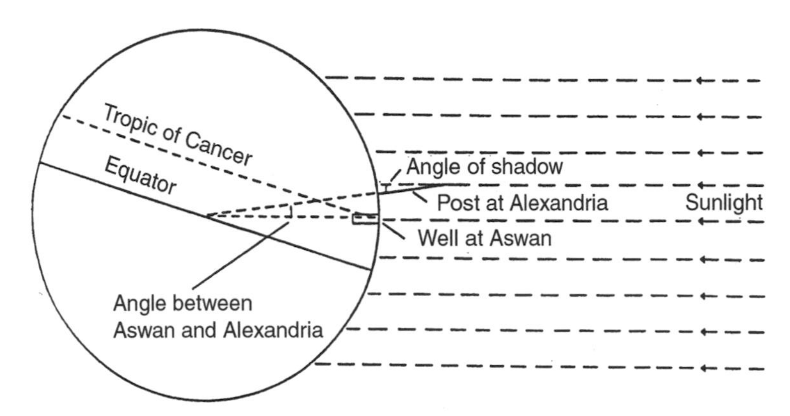


FIGURE 7.1 Geometry of Eratosthenes' method of measuring the circumference of the Earth.

time zone, and 24,000 miles for 24 time zones, which gets us back to New York again. You might protest that the arithmetic would be more accurate near the equator, where the time zones *really* are 1000 miles apart. If that bothers you, just deduct part of the flight time for takeoff and landing.

## GEOGRAPHY AND WEIGHT

Your weight depends on where you are on the Earth's surface. At the equator, the Earth is pulling you in, but the centrifugal force caused by the Earth's rotation is pulling you out radially. Let's calculate the magnitude of the effect and see if it is something to worry about.

The centrifugal field strength at the equator is

$$\omega^2 R = \frac{4\pi^2}{T^2} R = \frac{4\pi^2}{(24 \times 3600\ \mathrm{s})^2} (6.37 \times 10^6\ \mathrm{m})$$

$$= 0.034\ \mathrm{N/kg} \approx \frac{1}{3}\%\ g.$$

For a 70 kg person, the loss of weight at the equator due to centrifugal force amounts to about 200 g. Evidently the effect cannot compete with dieting.

Of course, at the North Pole there is no centrifugal force, but a person's weight is reduced there by the geometry of the Earth. The Earth is an oblate spheroid, with a belt around the equator, caused by centrifugal force during its formation. The radius to the North Pole is about 14 miles shorter than the mean radius of the Earth. One might expect, therefore, that the gravitational field at the pole would be larger by a factor of

$$\left(\frac{4000}{3986}\right)^2 = 1.007 = 0.7\% \text{ increase.}$$

The actual increase is only 0.18%. Although the distance from pole to equator is less than the mean radius, part of the mass of the Earth is farther away because it is spread out along the equatorial girdle.

A good approximation for g at sea level is $g = 9.7805\,(1 + 0.00529 \sin^2 \phi)$ where $\phi$ is the latitude. At the pole, $g = 9.8322$ N/kg, and at the equator $g = 9.7805$ N/kg.

## $g(R)$

The gravitational field strength on the Earth's surface, $g(R)$, is 9.8 N/kg. This is also called the acceleration due to gravity of an object in free fall. The units then are those of acceleration: m/s$^2$. These apparently different units are the same:

$$\frac{\text{N}}{\text{kg}} = \frac{(\text{kg})(\text{m/s}^2)}{\text{kg}} = \frac{\text{m}}{\text{s}^2}.$$

The magnitude of $g$ is $GM_{\text{Earth}}/R^2$

Are the astronauts in orbit in a much weaker field than they are when on Earth? For a close Earth orbit, where $r = R_{\text{Earth}} + \Delta r$, consider this approximation:

$$\frac{(R + \Delta r)^2}{R^2} = \frac{R^2 + 2R\,\Delta r + \Delta r^2}{R^2} \approx 1 + 2\frac{\Delta r}{R}.$$

Thus, a 1% error in $r$ yields a 2% error in $r^2$. For an Earth orbit that is 100 km above the surface, $\frac{\Delta r}{R} = \frac{1\times 10^5\,\text{m}}{6.4\times 10^6\,\text{m}} = 0.016$. The fractional increase in $r^2$ is twice this, and so is the fractional decrease in $g$. Therefore,

$$g(100\text{ km}) = 9.8(1 - 0.032) = 9.5\text{ N/kg}.$$

The astronauts in near-Earth orbit are evidently not "weightless" because of their distance from Earth.

The radius of the "hovering" or "geosynchronous" orbit is $4.2 \times 10^7$ m (which is 22,400 miles *above the Earth's surface*.) (See "Earth Orbits" later in this chapter). That's larger than the radius of the Earth by a factor of $\frac{4.2\times 10^7}{6.4\times 10^6} = 6.6$. Consequently, $g$ at that radius is smaller by a factor of $(6.6)^2 = 43$. Thus,

$$g\text{ (geosynchronous)} = 0.23\text{ N/kg}.$$

The Earth's field at the orbit of the Moon is very small. The radius is larger than the radius of the Earth by a factor of $\frac{240{,}000\text{ miles}}{4{,}000\text{ miles}} = 60$.

The Earth's field at that distance is smaller than 9.8 by a factor of 3600. At the moon, $g = 0.0027$ N/kg. Still, the Earth's field at the moon is strong enough to hang onto it. That centripetal force is

$$F_{\text{E–M}} = g_{\text{E at M}} M_{\text{Moon}}$$
$$= (2.7 \times 10^{-3}\text{ N/kg})(7.3 \times 10^{22}\text{ kg}) = 2.0 \times 10^{20}\text{ N}.$$

## HEIGHT OF ATMOSPHERE

How high is the sky? Or rather, what is the thickness of the atmosphere? As a first approximation, assume that the density of air stays constant as the height increases and the pressure decreases. The density of air at STP is 1.3 kg/m$^3$. (See "Density of Air" in chapter 9.) The atmospheric pressure at sea level is $1 \times 10^5$ N/m$^2$. To supply this pressure we need a column of air with height $h$:

$$\rho\, h\, g = P \quad \Rightarrow \quad (1.3)h\,(10) = 1 \times 10^5$$

$$h = 8 \times 10^3 \text{ m} = 8 \text{ km} \approx 5 \text{ miles.}$$

That's too small, even for a first approximation. We know that jet planes can fly at heights greater than 7 miles.

For a next approximation, assume that the air density stays constant for the first 4 km. Then, for the distance above the 4 km height, assume that the density is half that at sea level.

$$(0.65 \text{ kg/m}^3)h\,(10 \text{ N/kg}) = 5 \times 10^4 \text{ N/m}^2$$

$$h = 8\text{km}$$

With this second approximation we stretch the height of the atmosphere to a total of 12 km. We could keep going with the same procedure, but let's switch to a model that assumes density is proportional to pressure and therefore decreases continuously with height.

We will assume that the density is proportional to the pressure. This assumption is equivalent to assuming a constant temperature throughout the atmosphere—which is not a bad approximation:

$$\rho = \frac{P}{P_o}\rho_o \text{ . At sea level, where } P = P_o,\ \rho = \rho_o.$$

An increment of height of the column of air leads to an increment in pressure at the base. However, the pressure at the height $h + \mathrm{d}h$ is less than at the height $h$, and so we need a negative sign. Thus,

$$\mathrm{d}P = -\frac{P}{P_\mathrm{o}}\rho_\mathrm{o} g\ \mathrm{d}h \quad \Rightarrow \quad \frac{\mathrm{d}P}{P} = -\frac{\rho_\mathrm{o}}{P_\mathrm{o}} g\ \mathrm{d}h.$$

Let's use this approximation to calculate the pressure at a height of 12 km:

$$\frac{P}{P_\mathrm{o}} = \mathrm{e}^{-\frac{1.3}{10^5}(9.8)(1.2\times 10^4)} = \mathrm{e}^{-1.53} = 0.22.$$

At a height of 12 km, about a fifth of the atmosphere is still left. Compare this with our earlier 2-stage approximation, which called for no atmosphere at 12 km.

Let's find the height where only 10% of the atmosphere is left:

$$\frac{P}{P_\mathrm{o}} = 0.1 = \mathrm{e}^{-\frac{1.3}{10^5}(9.8)h} = \mathrm{e}^{-(1.27\times 10^{-4})h}.$$

$$\ln 0.1 = -1.27 \times 10^{-4}\ h \Rightarrow h = 18\ \mathrm{km} = 11\ \mathrm{miles}.$$

At this height and beyond, our simple approximations are no longer valid within 10%.

## DEPTH OF EARTH'S GRAVITY FIELD

Here on the surface of the Earth, we are in a gravity well. To see how deeply we are bound, let's put in the numbers for humans and for molecules. The gravitational potential is

$$V_{\text{grav. pot.}} = -G\frac{M}{R}$$

$$= -(6.7 \times 10^{-11})\,\frac{(6.0 \times 10^{24})}{(6.4 \times 10^{6})} = -6.3 \times 10^{7}\ \mathrm{J/kg}.$$

Therefore, a 70 kg human is bound by

$$U_{\text{grav}} = mV_{\text{grav pot}} = (70\text{ kg})(-6.3 \times 10^7\text{ J/kg}) = -4.4 \times 10^9\text{ J}.$$

That's about the energy contained in 44 gallons of gasoline and is equal to 1200 kW·hr.

On the other hand, an oxygen molecule is bound by

$$\begin{aligned} U_{\text{grav}} &= (32 \times 1.7 \times 10^{-27}\text{ kg})(-6.3 \times 10^7\text{ J/kg}) \\ &= -3.4 \times 10^{-18}\text{ J} = -21\text{ eV}. \end{aligned}$$

The gravitational binding energy of a helium atom is just one-eighth of this, or about 3 eV. Both of these energies are much larger than the average thermal energy of gas molecules in the atmosphere. (This is $^1\!/_{25}$ eV; see "Units and Approximations" at the front of the book.) Note, however, that helium escapes the Earth. (See "Moral of Tail" in chapter 4.) Fortunately, most of the oxygen doesn't.

## MOUNTAIN HEIGHT

How high can a mountain be on the Earth? The compressibility limit for granite is about 30,000 lbs/in$^2$ = $2 \times 10^8$ N/m$^2$. (About 1/100 of Young's modulus for steel.) At that pressure, the base rocks would flow rather than crumble. Let's calculate how high a mountain must be to exert the compressibility limit at its bottom.

The pressure exerted on its base by a mountain with a height $h$ is $\rho g h$. If that equals $2 \times 10^8$ N/m$^2$, then

$$h = \frac{2 \times 10^8}{(3 \times 10^3\text{ kg/m}^3)(9.8\text{ N/kg})} \approx 7\text{ km} \approx 4\text{ miles}.$$

Mount Everest is less than 6 miles high.

Another way of estimating the maximum height of a mountain is to picture a column of height $h$. The energy to melt the bottom meter of this granite column must be provided by the lowering of one meter of rock from the top to the base.

$$m\, g\, h = m\, L_{\text{fusion}}$$

$$h = \frac{L_{\text{fusion}}}{\text{g}} \quad \Rightarrow \quad \frac{2.4 \times 10^5 \text{ J/kg}}{10 \text{ N/kg}} = 24 \text{ km} = 15 \text{ miles.}$$

Here we used the value for the latent heat of fusion of silicon dioxide. The pressure at the base would not be uniform because of the crumbling, and pockets of molten rock would form when heights are smaller than we calculated.

The mountains on the Moon ought to be higher by a factor of 6 (about 35 miles) since the gravitational field strength on the surface of the moon is about one-sixth that of the Earth value. However, there has to be some process to lift the mountain up in the first place. On Earth we have continental movements to thrust up the mountains. Then, over a span of millions of years, they are eroded down. However, the Moon has long been seismically dead. Any mountains that were produced or that were left over from the formation have long since been reduced by continual bombardment of small and large objects from space. For this reason, the highest mountain on the moon is about 8 km. The mountains may have been thrown up by the collision of asteroids during the formation of the Earth-Moon system.

Mars has a gravitational field strength at the surface of

$$G\frac{M}{R^2} = 6.7 \times 10^{-11}\, \frac{6.4 \times 10^{23}}{(3.4 \times 10^6)^2} = 3.7 \text{ N/kg},$$

about 1/3 that of Earth. Mars has been active seismically within geological times. Its highest mountain is Olympus Mons, which is 15 miles high, almost three times the height of Everest.

# EARTH ORBITS

We have many Earth satellites that hover over the same place on the Earth's surface. They are in geosynchronous orbit, which has a period of 24 hours. How far away are they?

$$\text{centripetal force required} = \text{gravitational attraction}$$

$$\frac{mv^2}{r} = \frac{m\,4\pi^2 r}{T^2} = G\frac{mM}{r^2}$$

$$r^3 = \frac{GM}{4\pi^2}T^2 \text{ (Kepler's third law).}$$

For $T = 24\text{ hr} = 8.6 \times 10^4\text{ s}$,

$$r^3 = \frac{(6.7 \times 10^{-11})(6 \times 10^{24})}{4\pi^2}(8.6 \times 10^4)^2 = 7.5 \times 10^{22}$$

$$r = 4.2 \times 10^7\text{ m}$$

$$r_{\text{above surface}} = 4.2 \times 10^7 - 0.6 \times 10^7$$

$$= 3.6 \times 10^7\text{ m} = 22,400\text{ miles.}$$

A signal relayed up to this satellite and down again takes $\frac{7.2\times10^7\text{ m}}{3\times10^8\text{ m/s}} = 0.2$ s, creating a noticeable delay when two people are talking by way of the satellite.

The hovering satellites are not distributed evenly around the circumference at that altitude. There are three stable positions, separated by 120°. One of these is over the bulge of Brazil, and it is toward that satellite that TV disk receivers in the United States are aimed.

For calculations involving Earth at surface level, note that

$$mg = G\frac{mM}{R^2} \Rightarrow g = \frac{GM}{r^2} = \frac{(6.7 \times 10^{-11})(6 \times 10^{24})}{(6.4 \times 10^6)^2}$$

$$= 9.8\text{ N/kg (or m/s}^2\text{).}$$

For a satellite in orbit close to the surface of the Earth (or within a hundred miles or so),

$$T^2 = \frac{4\pi^2}{GM}R^3 = \frac{4\pi^2}{g}R \approx 4R \quad \Rightarrow \quad T = 2\sqrt{R} = 5000 \text{ s} = 1.4 \text{ hr}.$$

That was the period of Sputnik and the other early satellites.

## ESCAPE ENERGY FROM EARTH

Is NASA wasting our money by sending rockets away from the Earth?

The escape energy per kilogram from Earth is

$$E_{\text{energy per kg}} = G\frac{(1 \text{ kg})M_{\text{Earth}}}{R_{\text{Earth radius}}} = (1 \text{ kg})g\,R$$
$$= (1)(10)(6.4 \times 10^6 \text{ m}) = 6 \times 10^7 \text{ J}.$$

For comparison: 1 kW·hr = $3.6 \times 10^6$ J, costs about 15 cents

1 gal gas $\sim 1 \times 10^8$ J, costs about \$1.50 or about 40 cents/kg.

This, $6 \times 10^7$ J, electricity costs \$3.00 and gasoline costs \$2.50.

NASA pays about \$20,000 per kilogram to get to Earth orbit, and that requires only one-half the energy to escape completely. Why not just use a long extension cord and save money? (Note that we have used the identity: $g = G\frac{M}{R^2}$.)

## PRECESSION OF THE EQUINOXES

Polaris is our North Star only for now. Our Earth precesses like a giant top. The Sun exerts a torque on the Earth, trying to straighten it up so that its rotation plane will be the same as its orbital plane around the sun. Instead of straightening the spin plane, a torque makes a spinning top precess. Of course, there would be no torque if the Earth were spherical or if its plane of spinning were already in the orbital plane. However, the Earth has an equatorial girdle, caused by centrifugal forces during its formation. For ease of calculation, we will approximate the girdle as a ring with a thickness of 21 km, which is the difference between the equatorial and polar radius. We assume that the width of this ring is 6000 km. Since the Earth's radius is 6000 km, the ring in our model extends 30° on either side of the equator. The Earth's spin plane tilts at an angle of 23.5° from the orbital plane. As shown in the diagram, one side of the girdle is closer to the sun than the other side. The near side of the girdle is pulled toward the sun with greater force than the far side, thus creating a net torque.

Here are the constants and their values that we must use in this calculation:

$$M_{\text{Sun}} = 1.99 \times 10^{30}\ \text{kg}\,, \quad m_{\text{Earth}} = 5.98 \times 10^{24}\ \text{kg},$$

$$m_{\text{girdle}} = 1 \times 10^{22}\ \text{kg (calculated from the volume and}$$

$$\text{a density of } 2.5 \times 10^{3}\ \text{kg/m}^3)\,, \quad r_{\text{E-S}} = 1.5 \times 10^{11}\ \text{m},$$

$$R_{\text{Earth}} = 6.37 \times 10^{6}\ \text{m}\,, \quad G = 6.67 \times 10^{-11},$$

$$\omega = 2\pi/(24 \times 3600) = 7.3 \times 10^{-5}\ \text{radians/s}\,, \quad \theta = 23.5°,$$

For the Earth's moment of inertia use: $I = (0.34) m_{\text{E}} R_{\text{E}}^2$. (The number, 0.34, replaces 0.4 for the moment of inertia of a homogeneous sphere. The Earth is more dense near the center.)

The forces on the girdle are shown in figure 7.2. We assume that only a fourth of the girdle is responsible for each of the forces, and calculate their magnitude assuming that the mass of each quadrant of the girdle is located at a point. The middle two quadrants are ignored because their effects would cancel. The difference in forces on the two quadrants of the girdle is

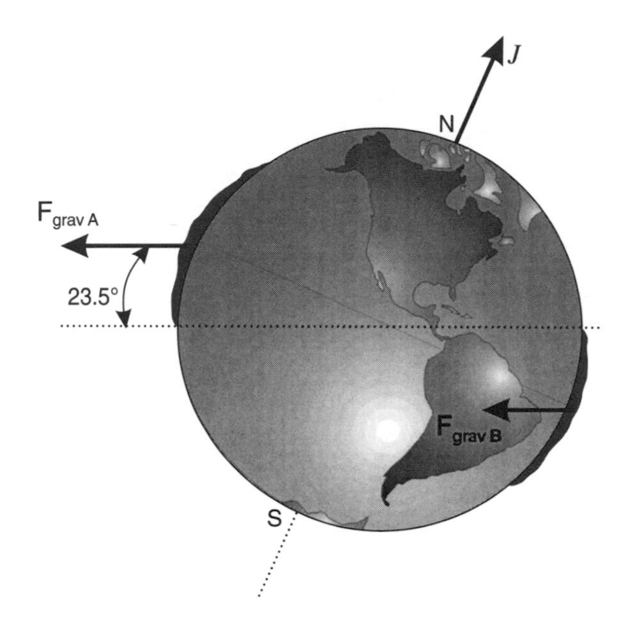


$$F_{\text{grav A}} > F_{\text{grav B}}$$

FIGURE 7.2 Approximate geometry of Earth's girdle, which is responsible for precession.

$$F_1 - F_2 =$$

$$G\left(\frac{1}{4}m_{\text{girdle}}\right)M_{\text{Sun}}\left[\frac{1}{(r_{\text{E-S}} - R\cos\theta)^2} - \frac{1}{(r_{\text{E-S}} + R\cos\theta)^2}\right]$$

$$F_1 - F_2 = Gm_{\text{girdle}}M_s\frac{R_E\cos\theta}{r_{\text{E-S}}^3}.$$

(Note the dependence on $\frac{1}{r^3}$, an indication of a dipole (or tidal) phenomenon.)

Precession frequency is the torque exerted on the spinning object divided by its angular momentum,

$$f_{\text{precession}} = \frac{\tau}{J} = \frac{(Gm_{\text{girdle}} M_{\text{S}} R_{\text{E}}^2 \cos\theta \sin\theta)/r_{\text{E-S}}^3}{(0.34) m_{\text{E}} R_{\text{E}}^2\, \omega}$$

$$= 1 \times 10^{-12}\ \text{rad/s}.$$

The actual precession frequency is $1.3 \times 10^{-12}$ rad/s. The precession period is the reciprocal of this: $T_{\text{precession period}} = \frac{1}{1.3\times 10^{-12}} = 7.7 \times 10^{11}$ s $= 24{,}400$ years. Even the pyramids are no longer aligned correctly!

## HOLE THROUGH THE EARTH

Starting out in New York, dig a hole through the center of the Earth and where do you end up? Not in China! Both New York and all of China are in the Northern Hemisphere. The antipode for New York is a few hundred miles to the west of Perth, Australia.

A standard physics problem is to find out what would happen if you could drop a cannon ball down a hole through the center of the Earth. We assume that the Earth has uniform density $\rho$ and the hole is along a diameter. Gauss's law tells us that the gravitational pull on an object in such a hole at radius, $r$, depends only on the mass of the sphere that has a radius smaller than $r$:

$$F(r) = -Gm\frac{\left(\frac{4}{3}\pi\, r^3\, \rho\right)}{r^2} = -m\left(\frac{4}{3}\pi\, G\, \rho\right) r.$$

Note that this condition produces a restoring force proportional to the negative of the displacement. That's the condition for *simple harmonic motion*. The period is

$$T = 2\pi\sqrt{\frac{m}{k}} = 2\pi\sqrt{\frac{1}{\frac{4}{3}\pi\, G\rho}}$$

$$= \sqrt{\frac{3\pi}{G\rho}} = \sqrt{\frac{3\pi}{(6.67 \times 10^{-11})(5500)}} = 5.1 \times 10^{3}\ \text{s} = 1.4\ \text{h}.$$

This period is the same as that of a low-altitude Earth satellite. (See "Earth Orbits" earlier in this chapter.) Note also that in the first expression, the period depends only on the value of $G$ and the density, but not on $r$! We don't have to bore a hole through the center of the Earth. A tiny marble would oscillate through a hole in a baseball if we could suspend the marble horizontally without friction. If the baseball had the same density as the Earth, the period for a round-trip of the marble would be 1.4 h. If instead we used a gold sphere—of any radius—the period would be 0.75 h.

## SLOWING OF THE EARTH'S ROTATION

You may not have noticed it, but the Earth is slowing down. The day is getting longer by 1.6 ms per century. This may not seem important enough to bother you, but the effect is enough to throw off by several hours the calculated eclipse times of 2000 years ago. To be sure, twenty centuries ago the day was shorter by only 32 ms, but the cumulative effect is surprising. The average day during these 2000 years was shorter by 16 ms. The cumulative shortening, day by day, was

$$(16 \times 10^{-3}\ \text{s/day})(2000\ \text{years})(365\ \text{days/yr})$$
$$= 1.17 \times 10^{4}\ \text{s} = 3.2\ \text{hrs}.$$

The Earth's rotation is slowing down because of tidal sloshing. The energy for all that wave and tidal motion has to come from some place: it is drained out of the rotational energy of the Earth. We cannot easily calculate the energy lost to tidal friction, but we can calculate the rotational energy lost as the Earth slows down. The energy of the spinning Earth is

$$\frac{1}{2} I\,\omega^2 = \frac{1}{2}(0.344)(5.98 \times 10^{24}\ \text{kg})(6.38 \times 10^6\ \text{m})^2 \left(\frac{2\pi}{24 \times 3600}\right)^2$$
$$= 22 \times 10^{28}\ \text{J}.$$

The *fractional* slowing of the Earth's rotation rate in one century is

$$\frac{1.6 \times 10^{-3}}{24 \times 3600} = 1.85 \times 10^{-8}.$$

The loss of rotational energy every century is

$$2 \times (1.85 \times 10^{-8})\,(22 \times 10^{28}\ \text{J}) = 8.1 \times 10^{21}\ \text{J}.$$

(The factor of 2 arises because the fractional slowing applies to the frequency $\omega$, which is squared in the energy formula.)

In one year, the Earth loses $8 \times 10^{19}$ J due to the tides. Compare this with the annual human consumption of energy of about $3 \times 10^{20}$ J, not including the food intake. This is ($5 \times 10^9$ people) ($1.5 \times 10^3$ kcal/day/person) (4200 J/kcal) (365 days/yr) $= 1.1 \times 10^{19}$ J!

But if the Earth slows down, it loses angular momentum. Somehow, the total angular momentum of the Earth-Moon system must be conserved. That can happen only if the angular momentum of the Moon in orbit around the Earth can appropriately increase. The Moon must move away from the Earth. The angular momentum of the Moon about the Earth is

$$I_M\,\omega_M = m R^2 \omega$$
$$= (7.34 \times 10^{22}\ \text{kg})(3.84 \times 10^8\ \text{m})^2 \frac{2\pi}{24 \times 3600 \times 27} = 29 \times 10^{33}.$$

The angular momentum of the Earth as it rotates is

$$I_E\,\omega_E = (0.344)(5.98 \times 10^{24})(6.38 \times 10^6)^2 \left(\frac{2\pi}{24 \times 3600}\right)$$
$$= 6.1 \times 10^{33}.$$

The loss of angular momentum of the Earth in one century is

$$(1.85 \times 10^{-8})(6.1 \times 10^{33}) = 1.1 \times 10^{26}.$$

(The first factor is the fractional change in $\omega$, which will produce the same fractional change in the Earth's momentum.)

The decrease in the angular momentum of the Earth from $L$ to $(L - \ell)$ is equal to the increase of the Moon's angular momentum as its orbital radius increases from $R$ to $(R + r)$. The fractional increase of the Moon's radius is $\frac{1}{2}(1.85 \times 10^{-8}) = 0.925 \times 10^{-8}$. (The factor of $\frac{1}{2}$ accounts for the fact that in the formula for angular momentum, $R$ is squared.) Therefore, the increase in radius is $(0.925 \times 10^{-8})(3.84 \times 10^8) = 3.6$ m/century.

Of course, if the Moon gets farther away from the Earth, it will climb out of the gravitational potential well. This process takes energy that must be transferred to the system by the tides. (The direct gravitational force is attractive.) The change in the Moon's potential energy is

$$\frac{GmM}{R} - \frac{GmM}{(R + r)} = \frac{GmM}{R}\,\frac{r}{R + r}$$
$$= 7.7 \times 10^{28}\,\frac{3.6}{3.84 \times 10^8} = 7.2 \times 10^{20}\ \text{J}.$$

The loss of Earth's rotational energy in one century is $8.1 \times 10^{21}$ J. It appears that most of the lost energy is dissipated in the tidal friction, and only about 9% is required to lift the Moon a small distance up its potential well.

# MASS OF EARTH'S GRAVITATIONAL FIELD

Energy is stored in gravitational, magnetic, and electric fields. But energy is mass is energy is mass: $E = mc^2$. One enormous reservoir of energy consists of the Earth's gravitational field. It must be massive! Does it contribute appreciably to the Earth's mass?

The *electric* field from a point or spherical source is

$$E = k\frac{q}{r^2} = \frac{1}{4\pi\varepsilon_0}\frac{q}{r^2}.$$

The energy density due to this field is

$$\mu_E = \frac{1}{2}\varepsilon_0 E^2 \text{ J/m}^2.$$

The parallel terms for the gravitational field produced by a sphere of mass $m$ are

$$g = G\frac{m}{r^2}.$$

However, the gravitational field produces binding energy, which is intrinsically negative. Therefore,

$$\mu_G = -\frac{1}{2}\frac{1}{4\pi G}g^2 = -\frac{G}{8\pi}\frac{m^2}{r^4}.$$

On the Earth's surface, the gravitational energy density is

$$-\frac{1}{8\pi G}g^2 = -\frac{1}{8\pi(6.7\times 10^{-11})}(9.8)^2 = -5.7\times 10^{10}\text{ J/m}^3$$

$$= -6.4\times 10^{-7}\text{ kg/m}^3 = -0.64\text{ mg/m}^3.$$

That's about one-billionth the density of water and about one-millionth the density of air. For comparison, consider the much smaller mass density of a very strong magnetic field—say, 10 teslas

(100,000 gauss):

$$\mu_B = \frac{1}{2}\frac{1}{\mu_0}B^2 = \frac{1}{2}\frac{1}{4\pi \times 10^{-7}}10^2 = 4.0 \times 10^7 \text{ J/m}^3$$

$$= 4.4 \times 10^{-10} \text{ kg/m}^3 = 4.4 \times 10^{-4} \text{ mg/m}^3.$$

For the entire Earth's field outside $r = R$, the gravitational mass-energy is

$$U = -\frac{Gm^2}{8\pi}\int_R^\infty 4\pi r^2 \frac{1}{r^4}\,\mathrm{d}r = -\frac{Gm^2}{2R}$$

$$= -1.9 \times 10^{32} \text{ J} = -2.1 \times 10^{15} \text{ kg}.$$

This is about one-billionth the mass of the Earth. The negative sign indicates that the Earth is missing this much mass—it is binding energy or binding mass.

We can account for the missing mass in terms of thermal radiation after the material crushed together. During the formation of the Earth, the material fell into its self-created potential well. The positive kinetic energy of the in rushing material increased steadily, and so did the negative potential energy. Their sum equaled zero. When the materials collided, the kinetic energy turned into internal thermal energy. As that radiated away, the sum of the mass-energy became more negative, which is the missing mass that we calculated. Actually, the mass of the Earth is continually increasing because of the radiation and particles we receive from the Sun.

Chapter

# 8

# Astronomy

## DIAMETER OF SUN AND MOON

In a classic children's book, *Many Moons* (1943), James Thurber asked the question, "How large is the Moon?" The next time there is a full Moon, stretch your hand out at arm's length and cover the Moon with the nail of your little finger. Evidently, the diameter of the Moon is less than the width of your fingernail! At least that's true in terms of angular measure.

To find the angular width of the Moon, measure the width of the fingernail on your little finger. Divide by the span from eye to fingernail when your arm is held straight out. That ratio of arc divided by radius is the angle in radians that your fingernail subtends:

$$\theta = \frac{\text{arc}}{r}.$$

To get the angle in degrees, multiply by 57°/rad (or just multiply by 60). Most people get about 1° for the angle subtended by their little finger. Since you can obscure the Moon with that fingernail, the Moon's angular width must be less than 1°. The actual width of the Moon is about $\frac{1}{2}^\circ \approx \frac{1}{120}$ radian.

Don't try the same method on the Sun! Besides, you already know that the angular width of the Sun must be about the same as that of the Moon: that's why we have total eclipses of the Sun.

The Moon is $3.8 \times 10^8$ m from the Earth, and the Sun is $1.5 \times 10^{11}$ m away. Therefore, the diameter of the Moon must be $d = \theta r = (1/120)(3.8 \times 10^8 \text{ m}) = 3.2 \times 10^6$ m (actually, $3.4 \times 10^6$ m). Similarly, the diameter of the Sun must be $(1/120)(1.5 \times 10^{11} \text{ m}) = 1.3 \times 10^9$ m (actually, $1.4 \times 10^9$ m).

You can get closer to the actual numbers by using $57° = 1$ rad, in which case you should multiply by $(1/114)$ instead of $(1/120)$.

A full Moon near the horizon, particularly a harvest Moon, appears much larger than when it is high up in the sky. It's an optical illusion. Against a background of trees, the Moon appears to be farther from the observer, and hence bigger. The sky itself appears to be not a hemisphere, but rather a flattened bowl, with stars that are overhead apparently closer than those on the horizon. You can break the Moon illusion by holding your little finger out at arm's length and observing that your finger can still obscure the Moon. Say "shrink" and the Moon will shrink before your eyes. Tbis exercise is best done when other people are not around!

## MINIMUM DISTANCE TO NEAREST STAR

Even our largest telescopes see stars only as point sources, not disks. Of course, we can see our Sun as a disk with an angular width of $1 \times 10^{-2}$ rad $\approx 1/2°$. One limitation of seeing stars as disks is caused by the limited angular resolution of any optical system. The optical resolution of a telescope with a mirror diameter of $d$ is given by $\theta = \frac{\lambda}{d}$. We might expect that the 200-inch telescope at Mount Palomar would have an angular resolution of $\theta = \frac{5 \times 10^{-7} \text{ m}}{5 \text{ m}} = 1 \times 10^{-7}$ rad. However, air turbulence reduces the resolution by more than a factor of 10. Even with larger telescopes at higher elevation in Hawaii, the limit of resolution is about 0.05 second of arc, which corresponds to $2.5 \times 10^{-7}$ rad.

If a star (the size of our Sun) were $4 \times 10^4$ times as far away as the Sun, it would subtend $\frac{1\times10^{-2}}{4\times10^4} = 2.5 \times 10^{-7}$ rad. In the best of our telescopes it would appear as a point source. Since we do not see any stars as disks, all the ones as large or larger than our Sun must be at least $4 \times 10^4$ times as far away from us as the Sun.

The distance from Earth to Sun is 8 light minutes $\approx$ 500 light seconds. Therefore, the minimum distance to such a star is $(4\times10^4)\times$ 500 light seconds $\approx$ ⅔ light years. (The distance to our nearest neighbor, Proxima Centauri, is about 4 light years.)

## PRESSURE OF SUNLIGHT

We are being bombarded, day and night, by sunlight. It continually exerts a force of repulsion between Earth and Sun. Will it knock us off course?

The radiation power of the Sun at the Earth's orbit is about 1.3 kW/m$^2$. If each photon carries 3 eV $= 5 \times 10^{-19}$ J, then the number of photons striking the Earth in one second in one square meter is

$$\frac{(1.3 \times 10^3\ \text{J/s}\cdot\text{m}^2)}{5 \times 10^{-19}\ \text{J/photon}} = 3 \times 10^{21}\ \text{photons/s}\cdot\text{m}^2.$$

The energy carried by a photon is $h\nu$ and the momentum it carries is $\frac{h\nu}{c}$. Since $F = \frac{\Delta\ \text{momentum}}{\Delta\ \text{time}}$ and $P = \frac{F}{A}$, the total pressure exerted by these photons is

$$P = \frac{(3 \times 10^{21})(5 \times 10^{-19})}{3 \times 10^8} = 5 \times 10^{-6}\ \text{N/m}^2.$$

That seems like a trivial amount, but the Earth's surface is large. Let's model the Earth as a flat disk with a radius of $6.4 \times 10^6$ m. Then the total force exerted on the Earth by sunlight is

$$F = (5 \times 10^{-6})\pi(6.4 \times 10^6)^2 = 6.4 \times 10^8\ \text{N}.$$

Since $10^4$ N, a metric ton, is about one American ton, the repulsive force of sunlight on the Earth is about 60,000 tons. Of course, that's very small compared with the gravitational attraction, which is

$$F_{\text{grav}} = G\frac{m_{\text{E}}M_{\text{S}}}{r_{\text{E-S}}^2} = (6.7 \times 10^{-11})\frac{(6 \times 10^{24})(2 \times 10^{30})}{(1.5 \times 10^{11})^2}$$
$$= 3.6 \times 10^{22}\text{ N} = 3.6 \times 10^{18}\text{ tons.}$$

## HOW MANY PHOTONS ARE REQUIRED TO SEE A FAINT STAR?

A dark-adapted eye, away from city lights, can see about 5000 stars at night. About 150 B.C. a Greek named Hipparchus, working at Rhodes and Alexandria, classified stars according to their brightness. He divided them into six groups, or magnitudes. As newer methods of measuring intensity have been developed during the last few centuries, the scales have always been adjusted to make them compatible with previous scales, which in turn trace back to Hipparchus. Consequently, the modern logarithmic scale of stellar intensities has five magnitudes for a change in intensity of 100. This leads to a log scale with a base of $\sqrt[5]{100} = 2.51$. Furthermore, so as not to offend Hipparchus or all the catalogs based on his work, the scale is reversed so that large magnitudes correspond to small brightness. The reference level is also historical and arbitrary, but on that scale the dimmest star that we can see with the unaided eye has magnitude +6; the brightest star, Sirius, has magnitude −1.5; and the Sun has magnitude −26.9. The formula for the difference between magnitudes in terms of intensities is

$$M_b - M_a = \log_{2.51}\left(\frac{I_a}{I_b}\right) = 2.5\log_{10}\left(\frac{I_a}{I_b}\right).$$

The stellar magnitude of the Sun = −26.9; M of the weakest visible star = +6; therefore, $\Delta M = 32.9$.

$$\Delta M = 2.5 \log_{10} \frac{I_1}{I_2} = 32.9 \quad \Rightarrow \quad \frac{I_1}{I_2} = 1.4 \times 10^{13}.$$

The radiation from the Sun at Earth's surface $\approx 1\ \text{kW/m}^2 = 1 \times 10^{-1}\ \text{J/s} \cdot \text{cm}^2$.

The radiation in the visible at Earth's surface $= 1 \times 10^{-2}\ \text{J/s} \cdot \text{cm}^2$.

The radiation from the faint star is $\frac{1\times10^{-2}}{1.4\times10^{13}} = 7 \times 10^{-16}\ \text{J/s} \cdot \text{cm}^2$.

The photon radiation is $\frac{7\times10^{-16}\ \text{J/s}\cdot\text{cm}^2}{3\times10^{-19}\ \text{J/photon}} = 2 \times 10^3\ \text{photons/s} \cdot \text{cm}^2$.

Multiply by $1/3$ for area of dark adapted pupil, by $1/10$ for image retention time, and by $1/10$ for reaching the retina after reflection and absorption in the eye.

It takes about 7 photons in one retention period to detect a star, and thus about 70 photons per second.

## FUELING THE SUN

These days everyone knows that the stars are fueled by nuclear reactions, fusing hydrogen into helium. But couldn't some chemical reaction do just as well? We can calculate the total power produced by the Sun because we know the power intercepted by the Earth. At the top of the atmosphere it is 1.3 kW/m$^2$. A spherical shell around the Sun, with a radius equal to the orbital radius of the Earth, would intercept

$$[4\pi(1.5 \times 10^{11})^2\ \text{m}^2](1.3 \times 10^3\ \text{W/m}^2) = 3.7 \times 10^{26}\ \text{J/s}.$$

If the Sun derived its energy from chemical reactions, we might get several eV per molecular interaction. For a mole of such molecules, we would get ($6 \times 10^{23}$ particles per mole) $\times$ (3 eV per interaction)

$= 2 \times 10^{24}$ eV per mole $= 3 \times 10^5$ J per mole. Assume that a mole of such fuel (perhaps a mix of gasoline and oxygen) has a mass of 20 g. Then a tenth the mass of the Sun would contain

$$\left(\frac{2 \times 10^{29}\ \mathrm{kg}}{0.02\ \mathrm{kg/mole}}\right) = 1 \times 10^{31}\mathrm{moles}.$$

The energy produced by burning a tenth of the sun would be

$$(3 \times 10^5\ \mathrm{J/mole})(1 \times 10^{31}\ \mathrm{moles}) = 3 \times 10^{36}\ \mathrm{J}.$$

Since the Sun produces $3.7 \times 10^{26}$ J/s, the chemical fires using a tenth of the Sun's mass would last about $10^{10}$ s. That's about 300 years. Evidently, chemical burning isn't sufficent to fuel the Sun.

Perhaps the Sun still has the energy that was released when the cosmic dust gathered to form the star. That certainly got the center hot enough to trigger the nuclear fusion. For an order of magnitude calculation, assume that every kilogram fell from infinity to the fully formed Sun. Of course, the early-comers didn't lose so much energy during the in-fall because the total mass hadn't gathered yet.

The potential energy of the mass of the Sun after it falls from infinity to the radius of the Sun is

$$E_{\mathrm{grav}} = -G\frac{M^2}{R} \approx -10^{-10}\frac{10^{60}}{10^9} = -10^{41}\ \mathrm{J}.$$

Since the Sun produces $4 \times 10^{26}$ J/s, the gravitational energy released could have provided the energy radiated for

$$\frac{10^{41}\ \mathrm{J}}{4 \times 10^{26}\ \mathrm{J}/s} = 2 \times 10^{14}\ \mathrm{s} = 10^7\ \mathrm{years}.$$

That's a lot of energy, and it is enough to start the fusion process, but our Sun has been glowing for almost $10^{10}$ years.

# AGE OF THE ELEMENTS

Our solar system, and everything in it except for hydrogen and helium, was formed out of the debris of a nova. We are literally made of star dust. When the imploding star cooked the heavier elements, it produced many radioactive isotopes that have since decayed away. In table 8.1 are listed the radioactive isotopes that yet remain, except for those that are products of other decays. $C^{14}$ is not listed because it is continually being generated by neutron bombardment of nitrogen at the top of the atmosphere.

If we assume that isotopes that are close in the number of neutrons were made in equal quantities, then it appears that the nova must have occurred more than a billion years ago. The original radioactive elements with half-lives smaller than a billion years are mostly gone.

The current ratio of the abundance of $U^{238}$ to $U^{235}$ is about 140. If the age of the elements is $9 \times 10^9$ years, then the amount of $U^{238}$ would have decreased by 4. The age of $U^{235}$ would be equal to $\frac{9}{0.71} = 12.7$ half-lives, and thc amount would have decreased by about a factor of 6700. In that case, the ratio of $U^{238}$ to $U^{235}$ would be about 1700.

If the age of the elements is $6 \times 10^9$, then the amount of $U^{238}$ would have decreased by 2.5. The amount of $U^{235}$ would have decreased during $\frac{6.0}{0.71} = 8.5$ half-lives, or by about a factor of 360. In that case,

**Table 8.1**
**Naturally Occurring Radioactive Isotopes on the Earth**

| *Element* | *Half-life in years* |
|---|---|
| ${}_{92}U^{238}$ | $4.5 \times 10^9$ |
| ${}_{92}U^{235}$ | $7.1 \times 10^8$ |
| ${}_{90}Th^{232}$ | $1.4 \times 10^{10}$ |
| ${}_{71}Lu^{176}$ | $1.0 \times 10^{10}$ |
| ${}_{37}Rb^{87}$ | $5.0 \times 10^{10}$ |
| ${}_{19}K^{40}$ | $1.3 \times 10^9$ |

the ratio of $U^{238}$ to $U^{235}$ would be about 145, which is very close to the observed ratio.

Here is a direct method of finding the age of the elements from this data. If the $U^{238}$ has undergone $n$ half-lives, then $U^{235}$ has undergone $\frac{4.5\times10^9}{7.1\times10^8}n = 6.3n$ half-lives. Although the two isotopes started out with equal quantities, the $U^{238}$ is now 140 times as plentiful.

$$140 = \frac{2^{-n}}{2^{-6.3n}} = 2^{5.3n} = e^{0.69(5.3n)} = e^{3.7n}$$

$$\ln 140 = 4.9 = 3.7n, \quad n = 1.3.$$

The life-time of the elements is 1.3 times the half-life of $U^{238}$, and so is equal to 6 billion years, which is the value we deduced by the first method.

## THE 21 CM LINE

The universe is filled with hydrogen atoms left over from the Big Bang. They form a spherical cloud around our galaxy, with an average density of close to one per cubic centimeter. The hydrogen atom has a very low-lying excited state caused by a coupling between the magnetic moments of the proton nucleus and the electron. In the ground state the magnetic moments are aligned. A collision with another atom can flip the relative alignment, raising the energy of the system by $5.9 \times 10^{-6}$ eV. Because the energy is so small, the relaxation time is very long—about a million years. When the system does flip back, it emits a signal with a frequency of 1428 MHz and a wavelength of 21 cm. This wavelength is about twice that in a kitchen microwave oven. The signal is so distinctive that it can be used to map galactic structure.

We can appeal to a classical model to get a good approximation to the excitation and energy level of this phenomenon. The potential energy of two magnetic dipoles aligned along their axes is

$$U = \frac{\mu_0}{2\pi}\frac{\mathbf{m}_1\mathbf{m}_2}{r^3}.$$

The energy absorbed when the proton and electron are flipped out of line is

$$\Delta E = \frac{2\mu_0}{2\pi}\frac{\mathbf{m}_\mathrm{p}\mathbf{m}_\mathrm{e}}{r^3} = (4 \times 10^{-7})\frac{(1.4 \times 10^{-26})(9.3 \times 10^{-24})}{(0.53 \times 10^{-10})^3}$$
$$= 3.5 \times 10^{-25}\ \mathrm{J} = 2.2 \times 10^{-6}\ \mathrm{eV}.$$

When they flip back into line again, the emitted photon has a wavelength of

$$\lambda = \frac{h\,c}{(h\,\nu)} = \frac{(6.6 \times 10^{-34})(3 \times 10^{8})}{(3.5 \times 10^{-25})} = 57\ \mathrm{cm}.$$

This approximation yields at least the right order of magnitude. Note that we have used the Bohr radius and that the final answer is very sensitive to the value of $r$, which gets cubed. In the quantum model of the hydrogen atom, the possible position of the electron is distributed over a range of $r$. When the electron is at a smaller radius, the dipole coupling energy is larger and the wavelength of the emitted photon is shorter. We have also left out a correction factor that takes into account the three-dimensional nature of the dipole interaction. The dipoles can be not only in line with each other, but they can also be side by side.

## THE PRODIGAL SUN

The Sun emits a tremendous amount of energy every second. Since energy is mass, that means that the Sun is steadily losing mass. How much? The electromagnetic flux at the radius of the Earth is $1.2 \times 10^3\ \mathrm{J/s \cdot m^3}$. The flux through a spherical shell at that radius is

$$(1.2 \times 10^3\ \mathrm{J/s \cdot m^2})4\pi(1.5 \times 10^{11}\ \mathrm{m})^2 = 3.4 \times 10^{26}\ \mathrm{J/s}.$$

The mass radiated per second is $\frac{3.4\times10^{26}}{(3\times10^{8})^{2}} = 3.8 \times 10^{9}$ kg. That's about four million tons per second. In a year, the Sun radiates $\pi \times 10^{7}$ that much, or $1.2 \times 10^{17}$ kg.

Are we running out of Sun? If this loss has been going on at the same rate for five billion years (the age of the Earth and almost the age of the Sun), then the Sun has lost

$$(1.2 \times 10^{17} \text{ kg/yr})(5 \times 10^{9} \text{ yrs}) = 6 \times 10^{26} \text{ kg}.$$

All is not lost, however. The mass of the Sun is $2 \times 10^{30}$ kg. In five billion years the Sun has lost only 1 part in 3300 of its mass. Before the Sun runs out of mass, it (and we) will undergo far worse catastrophes.

Chapter

# 9

# Atoms and Molecules

## ALL ATOMS ARE (ABOUT) THE SAME SIZE

First, let's find the diameter of an aluminum atom. A mole ($6 \times 10^{23}$) of aluminum atoms has a mass of 27 g. (The atomic "weight" is A = 27 g/mole). The density is $\rho = 2.7\ \text{g/cm}^3$.

$$\frac{A}{\rho} = \frac{27\ \text{g/mole}}{2.7\ \text{g/cm}^3} = 10 \frac{\text{cm}^3}{\text{mole}}.$$

A cube with a volume of 10 $\text{cm}^3$ has a side of about 2 cm (2.15 cm). The cube contains $6 \times 10^{23}$ atoms. Since $6 \times 10^{23} \approx 10^{24}$, the cube root is about $10^8$ ($8.4 \times 10^7$). Therefore, in solid aluminum, $10^8$ atoms are lined up on a side that is 2 cm long. The diameter of an aluminum atom must be $2 \times 10^{-8}$ cm. Or, to two significant figures:

$$\frac{2.15\ \text{cm/side}}{8.4 \times 10^7\ \text{atoms/side}} = 2.6 \times 10^{-8}\ \text{cm/atom}.$$

One can use the same method to find the molar volume and the atomic diameter of a range of elements in the periodic table (see table 9.1).

## Table 9.1
## Molar Volume and Atomic Diameter of Representative Elements in the Periodic Table

| *Element* | *Z* | *A* | *ρ (g/cm³)* | *Volume (cm³)* | *Atomic Diameter ($10^{-8}$cm)* |
|---|---|---|---|---|---|
| H (solid) | 1 | 1 | 0.076 | 13.0 | 2.6 |
| Be | 4 | 9 | 1.8 | 5.0 | 2.0 |
| C | 6 | 12 | 2.3 | 5.2 | 2.1 |
| Na | 11 | 23 | 0.97 | 24.0 | 3.4 |
| Fe | 26 | 56 | 7.9 | 7.1 | 2.3 |
| Au | 79 | 197 | 19.3 | 10.0 | 2.6 |
| Pb | 82 | 207 | 11.3 | 18.0 | 3.1 |
| U | 92 | 238 | 19.1 | 12.5 | 2.8 |

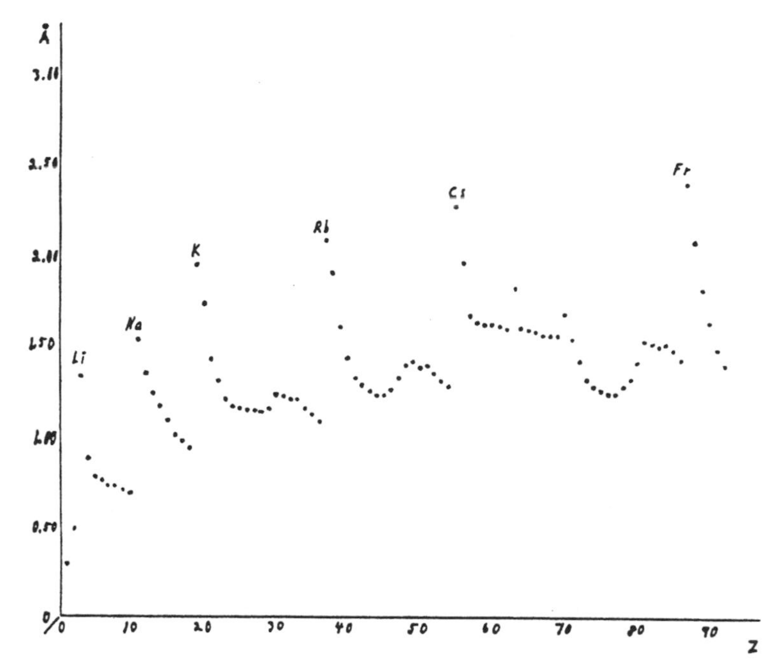


FIGURE 9.1 Atomic radii (single-bond covalent) as function of atomic number.

Note the following:

1. A mole of any solid (or liquid) element has a volume of 10 $cm^3$ within a factor of 2. Compare this with the volume of a mole of ideal gas at STP, which is 22.4 liters. = 22,400 $cm^3$.
2. The radius of an atom is $1 \times 10^{-8}$ cm (1 angstrom) within a factor of 2. This is a clue to the structure of atoms. As more electrons are added for the elements higher in the periodic table, the inner electrons must be drawn in so that the size of the atom remains about constant. Figure 9.1 shows the variations of the radius of free atoms as a function of Z. The alkali atoms are large because the single valence electron is not tightly held.
3. The calculations here are for the size when bound. An isolated hydrogen atom has a diameter of only $1.06 \times 10^{-10}$ m. This size can be calculated from the ground state of the Bohr orbits (see the note on the 21 cm line).
4. Considering its place in the periodic table, lead is very undense.

## DENSITY OF AIR

Unless air is moving past us very rapidly, we tend to think of it as "airy nothing." Take a guess at the mass of the air in your room. Now let's calculate using the following information.

Molecular "weight" for air $= A = 29$ g/mole.

Volume of 1 mole of ideal gas at STP $= 22.4\ell = 2.24 \times 10^{-2}\ \text{m}^3$.

Density of air at STP $= 29 \frac{\text{g}}{\text{mole}} \times \frac{\text{mole}}{2.24 \times 10^{-2}\ \text{m}^3} = 1.3 \times 10^3\ \text{g/m}^3 = 1.3\ \text{kg/m}^3$

What is the mass of the air in a room 10 ft × 20 ft × 8 ft? (Volume of room $= 1600\ \text{ft}^3 = 45\ \text{m}^3$, mass of air in room $= (45\ \text{m}^3)(1.3\ \text{kg/m}^3) = 60$ kg.) No offense, but that's probably almost as much as you weigh!

# MOLECULAR SPACING IN A GAS

In many elementary school science texts, the atoms in solids are portrayed as little circles, lined up shoulder to shoulder. The gas circles are shown, separated and dashing around. Unfortunately, the circles for liquids are drawn to be somewhat in-between, moving around with considerable space between each one. Of course, the atoms in a liquid must be touching one another, since material in the liquid state has about the same density as the solid material.

But how far apart should those gas atoms be? The volume of 1 mole of ideal gas at STP $= 22.4\ell = 2.24 \times 10^{-2}$ m$^3$. The volume of a mole of solid $\approx 1 \times 10^{-5}$ m$^3$ (See "All Atoms Are (about) the Same Size" at the beginning of this chapter.)

$$\frac{\text{Volume 1 mole gas}}{\text{Volume 1 mole solid}} = \frac{2 \times 10^{-2}}{1 \times 10^{-5}} = 2000.$$

Therefore, the average spacing of gas atoms $= \sqrt[3]{2000} = 13$ molecular diameters.

# MOLAR ENERGY

Each molecular energy exchange involves about 3 eV $= 5 \times 10^{-19}$ J. A mole of such interactions would involve

$$(6 \times 10^{23} \text{ interactions/mole})(5 \times 10^{-19} \text{ J/interaction})$$
$$= 3 \times 10^5 \text{ J/mole}$$

All hydrocarbons yield about the same energy per mass. A mole of $CH_4$ (methane) has a mass of 16 g. Each molecule of $CH_4$ would combine with oxygen to produce $CO_2$ and $H_2O$, each reaction yielding 3 eV;

$$CH_4 + 2O_2 \rightarrow CO_2 + 2H_2O.$$

**Table 9.2**
**Values of Heats of Combusion for Hydrocarbons**

| *Name* | *Formula* | *Molecular Weight* | *J/mole* | $\frac{\text{J/mole}}{\text{molwght}} = \frac{\text{J}}{\text{gram}}$ |
|---|---|---|---|---|
| Methane | $CH_4$ | 16 | $8.9 \times 10^5$ | $56 \times 10^3$ |
| Acetylene | $(CH)_2$ | 26 | $1.3 \times 10^6$ | $50 \times 10^3$ |
| Amylene | $C_5H_{10}$ | 70 | $3.4 \times 10^6$ | $48 \times 10^3$ |
| Benzene | $C_6H_6$ | 78 | $3.3 \times 10^6$ | $42 \times 10^3$ |
| Camphene | $C_{10}H_{16}$ | 136 | $6.2 \times 10^6$ | $45 \times 10^3$ |
| Cycloheptene | $C_7H_{12}$ | 96 | $4.4 \times 10^6$ | $46 \times 10^3$ |
| Decane | $C_{10}H_{22}$ | 142 | $6.8 \times 10^6$ | $48 \times 10^3$ |
| N-octane | $C_8H_{18}$ | 114 | $5.5 \times 10^6$ | $48 \times 10^3$ |

The burning of a mole of $CH_4$ would thus release $6 \times 10^5$ J. (The handbook value is closer to $9 \times 10^5$ J/mole.)

Table 9.2 shows the handbook values of heats of combustion for some hydrocarbons. Observe that all the hydrocarbons yield about the same energy per mass.

A gallon of gasoline (n-octane) has a mass of about 3 kg and when burned would provide $150 \times 10^6$ J, which matches the handbook value for hydrocarbons of $1.3 \times 10^8$ J/gal.

## LATENT HEAT OF FUSION AND VAPORIZATION

In elementary and junior high schools, a distinction is often made between physical and chemical changes. Such distinction is hard to define consistently. To be sure, "physical changes" usually involve smaller transformation energy than "chemical changes." Consider, however, the phase change energies (latent heats) of water fusion and vaporization. In calories per gram these are, respectively, 80 cal/g and 540 cal/g (at STP). For comparison with atomic energy values, change the units to eV.

$$80\frac{\text{cal}}{\text{g}} = 330\frac{\text{J}}{\text{g}}$$

$$= 6 \times 10^3 \frac{\text{J}}{\text{mole}} \left(\frac{6 \times 10^{18}\ \text{eV}}{1\text{J}}\right)\left(\frac{1\ \text{mole}}{6 \times 10^{23}\ \text{molecules}}\right)$$

$$= 0.06\ \text{eV/molecule} \approx \frac{1}{16}\ \text{eV/molecule}.$$

For evaporation (or boiling), $\Delta U = \frac{540}{80} 0.06\ \text{eV} = 0.4\ \text{eV}$. The value for water vaporization is greater than that for many chemical transitions, such as the breakup of sulfur bromide.

The latent heat of vaporization provides energy for two processes. The major share goes into the energy of breaking bonds that hold the molecules together in a liquid. The work of expanding the freed molecules into a gas at atmospheric pressure is

$$P\,\Delta V = (1 \times 10^5\ \text{N/m}^2)(22.4\ \ell)\left(\frac{1\ \text{m}^3}{1 \times 10^3\ \ell}\right)$$

$$= 2240\ \text{J} = 0.22 \times 10^4\ \text{J}.$$

Compare this work with the total latent heat of vaporization for a mole of water:

$$\left(540\frac{\text{cal}}{\text{g}}\right)\left(\frac{18\ \text{g}}{\text{mole}}\right)\left(\frac{4.2\ \text{J}}{1\ \text{cal}}\right) = 4.08 \times 10^4\ \text{J}.$$

About 5% of the latent heat of vaporization goes into the work of expansion.

While physicists use electron volts per atomic or particle event, chemists use joules or calories per mole. Knowing the relationship between the two values is useful when talking to chemists:

$$1\frac{\text{eV}}{\text{particle}}\left(\frac{1.6 \times 10^{-19}\ \text{J}}{1\ \text{eV}}\right)\left(\frac{6 \times 10^{23}\ \text{particles}}{1\ \text{mole}}\right)$$

$$= 1 \times 10^5\ \text{J/mole} \approx 25{,}000\ \text{cal/mole} = 25\ \text{kcal/mole}.$$

## SURFACE TENSION

Molecules on the surface between air and a liquid are pulled back toward the liquid, attracted by the molecules underneath. When the surface is streched, a horizontal force resists the expansion. Pulling a molecule to the surface, expanding the surface area, takes energy. We guess that this energy is some fraction, $\alpha$, of the heat of vaporization when all bonds are broken. For water, $\Delta U_{\text{vaporization}} = 0.4 \text{ eV} = 6.4 \times 10^{-20}$ J. (See the previous section.) The energy to break part of these bonds and move the molecule into the surface comes from the horizontal force of surface tension

$$\alpha\,\Delta U = F\,\Delta x \quad \Rightarrow \quad \alpha(6.4 \times 10^{-20}) = F\,(3 \times 10^{-10}).$$

(A molecule of water has a diameter of $3 \times 10^{-10}$ m.) In one meter there are $\frac{1}{3} \times 10^{10}$ molecules. Therefore,

$$F/\text{meter} = \frac{\alpha\,(6.4 \times 10^{-20})}{3 \times 10^{-10}} \left(\frac{1}{3} \times 10^{10}\right) = \alpha\,0.71\ \text{N/m}.$$

The handbook value for $\gamma$, the coefficient of surface tension for water, is 0.073 N/m. The result we obtained would require that $\alpha$ equal $\frac{1}{10}$.

Using the same assumptions for mercury, with a value of $\Delta U_{\text{vaporization}} = 9.2 \times 10^{-20}$ J, yields a $\gamma$ value for mercury of $1.1\alpha$ N/m. The handbook value of $\gamma$ for mercury is 0.48 N/m. In this case, $\alpha = 0.4$. This value is much closer to what we would expect for an atom freed from half its bonds in being pulled to the surface. The amount of energy needed to pull a water molecule into the surface is relatively small, indicating a complex surface effect.

## BINDING FORCE BETWEEN ATOMS

Each atom in a solid or liquid can be considered bound in a potential well formed by its immediate neighbors. The potential energy well must look something like the illustration in figure 9.2.

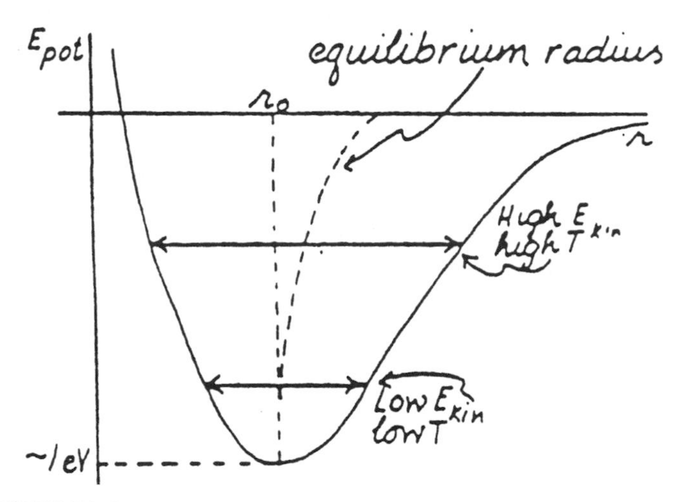


FIGURE 9.2 Potential energy well for bound atom.

The curve is asymmetric because if the atom gets too close to its neighbor it will be sharply repelled by the Pauli exclusion principle acting on the outer shell electrons. For radial distances beyond its equilibrium position, the potential energy curve has positive slope, corresponding to an attractive restoring force. That curve gets steeper up to the point where the radius is about twice the equilibrium radius. Then the slope decreases, corresponding to a weaker force. The effect is like stretching a spring until it is beyond its elastic limit.

Our model calls for the trapped atom to oscillate in the well, with its kinetic energy corresponding to the temperature. The fact that the potential well is asymmetric explains why solids and liquids expand when they get hotter. It is not a matter of the oscillations having greater amplitude, but rather the equilibrium radius gets larger at higher temperatures because of the asymmetry.

To complete the graph of $U(r)$, we must determine values for the axes. The equilibrium position is the atomic radius $\approx 1 \times 10^{-10}$ m. The strength of the binding (the depth of the well) differs depending on whether the condtions are ionic, covalent, metallic, or van der Waals. An example of almost pure ionic binding is NaCl. The valence electron (3s) of Na is almost completely captured by the Cl to form a complete shell (3p). That leaves a $Na^+$ ion adjacent to a $Cl^-$ ion. Their electrostatic attraction can be calculated by putting in a few numbers.

To free the electron from Na: +5.14 eV

The gain in binding energy by adding the electron to Cl: −3.61 eV.

The ionic electrostatic binding energy:

$$U = -k\frac{q_1 q_2}{r}$$

$$= -9 \times 10^9 \frac{(1.6 \times 10^{-19})^2}{2.36 \times 10^{-19}} \frac{1}{(1.6 \times 10^{-19}\ \mathrm{J/eV})} = -6.1\ \mathrm{eV}$$

The sum of these credits and debits is −4.6 eV. The handbook value of binding energy for NaCl is −4.25 eV. The transferred electron is not completely attached to the Cl, and so the binding is a little less than we calculated.

Covalent bonds occur when two atoms share one or two electrons. The best example is H-H, the hydrogen molecule, which is bound by 4.3 eV. The two electrons spend part of their time between the two protons. Carbon molecules also have bonds in the several eV range.

In metallic bonding, one or two valence electrons from each atom in a crystal are freed to wander throughout the crystal. They no longer belong to their parent atom. The remaining positive ions are bound

**Table 9.3**
**Atomic Binding Energies**

| *Substance* | *eV/atom* |
|---|---|
| Iron | 3.5 |
| Aluminum | 2.9 |
| Carbon (graphite) | 7.2 |
| Lead | 1.8 |
| Gold | 3.4 |
| Copper | 3.0 |
| Uranium | 4.6 |

together by the negative electrons interweaving among them. We can determine the binding energy of each atom by examining the vaporization energy of the material. In handbooks these are given in terms of kJ/mole (or in chemistry books, kCal/mole). Table 9.3 lists a number of these, converted to eV/atom. Note that most of these atoms are bound by a few eV, in agreement with other phenomena where molecules are broken up or rearranged by high temperatures or radiant energy. For instance, most electrochemical cells furnish a voltage between 1 and 2 volts. Also, blue photons with energies of about 3 eV can affect chemical bonds in our eyes, on photographic film, in photosynthesis, and in bleaching.

The van der Waals forces that bond molecules such as helium are much weaker than the other forces. They are caused by dipole-dipole interactions with a $\frac{1}{r^6}$ dependence.

For an approximate model of atomic bonding, we will assign a value of 3 eV ($5 \times 10^{-19}$ J) to the depth of the potential well. Then we can calculate the atomic binding force and the spring constant of the binding.

For an object in a potential well, the binding force is $F = -\frac{dU}{dr}$. The graph of $F(r)$ is derived from the slope of the curve of $U(r)$ (fig. 9.3). Note that at the equilibrium radius, $F(r_0) = 0$. For $r < r_0$, the force is positive, repulsion. For $r > r_0$, the force is negative, in the direction to bring the atom back to $r_0$.

In this case, the maximum attractive force over a range of 1 radius is

$$F = -\frac{\Delta U}{\Delta r} = \frac{5 \times 10^{-19}\ \text{J}}{1 \times 10^{-10}\ \text{m}} = 5 \times 10^{-9}\ \text{N}.$$

Is this a plausible value for atomic binding force? Calculate the force to rip apart a steel bar with a square cross section of 1 $m^2$. Along one edge of the bar, each atom occupies $2 \times 10^{-10}$ m. Therefore, there are $5 \times 10^9$ atoms along an edge, and $25 \times 10^{18}$ atoms in the cross section. If each exerts a binding force of $5 \times 10^{-9}$ N, the stress needed to pull the bar apart is

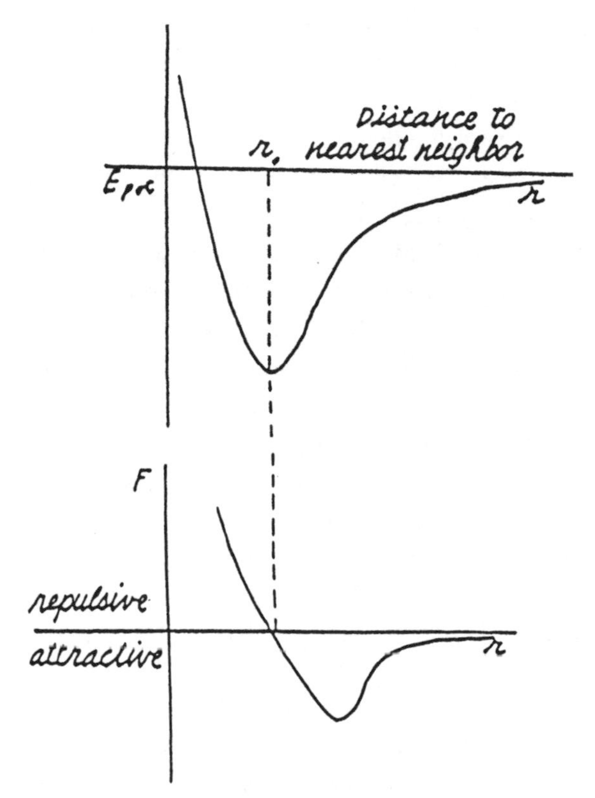


FIGURE 9.3 Potential energy and force on bound atom.

$$25 \times 10^{18} \text{ atoms/m}^2 \times 5 \times 10^{-9} \text{ N/atom} \approx 1 \times 10^{11} \text{ N/m}^2.$$

But this is close to Young's modulus for steel ($2 \times 10^{11}$) or any other hard metal! Stress is proportional to strain: $\frac{F}{A} = -Y\frac{\Delta L}{L}$, where $Y$ is Young's modulus. The approximations must be reasonably correct.

Yield pressure for steel is about $5 \times 10^9$ N/m². Note the significance.

$$\frac{\Delta L}{L} = -\frac{1}{Y}\frac{F}{A} = -\frac{1}{2 \times 10^{11}} 5 \times 10^9 = \frac{1}{40}.$$

Steel cannot stand elastic stretching or compression greater than 1/40 its length.

## ATOMIC SPRING CONSTANT

If the binding energy between two atoms is 3 eV, then the maximum binding force between the two atoms is $F = 5 \times 10^{-9}$ N. (See the previous section.) Modeling Hooke's law, we have

$$F = -k\,\mathrm{r} \quad \Rightarrow \quad k = \frac{5 \times 10^{-9}\ \mathrm{N}}{1 \times 10^{-10}\ \mathrm{m}} = 50\ \mathrm{N/m}.$$

(The minus sign disappears because $F$ is a restoring force, in the opposite direction from $r$.)

Let's apply this to the vibration and rotation of the hydrogen molecule.

$$f_{\mathrm{vibration}} = \frac{1}{2\pi}\sqrt{\frac{k}{m}} = \frac{1}{2\pi}\sqrt{\frac{50\ \mathrm{N/m}}{0.8 \times 10^{-27}\ \mathrm{kg}}} = 4 \times 10^{13}\ \mathrm{Hz}.$$

(The denominator is the "reduced" mass, since the oscillation is between two protons, each having mass of $1.7 \times 10^{-27}$ kg.) The energy of this oscillation is

$$\begin{aligned} E_{\mathrm{vibration}} &= h\nu = (6.6 \times 10^{-34}\ \mathrm{J \cdot s})(4 \times 10^{13}\ \mathrm{Hz}) \\ &= 3 \times 10^{-20}\ \mathrm{J} = 0.17\ \mathrm{eV}. \end{aligned}$$

The energy of rotation of the hydrogen molecule is

$$\begin{aligned} E_{\mathrm{rotation}} &= 1/2\,\frac{L^2}{I} = 1/2\,\frac{10^{-68}}{2(1.7 \times 10^{-27})(1 \times 10^{-10})^2} \\ &= 1.5 \times 10^{-22}\ \mathrm{J} = 1 \times 10^{-3}\ \mathrm{eV}. \end{aligned}$$

(The moment of inertia of the two protons is $2mr^2$. The smallest angular momentum the molecule can have is one unit—which is Planck's constant, $\hbar = h/2\pi = 1 \times 10^{-34}$ kg · m$^2$.)

Note that the oscillation frequency for the hydrogen molecule is in the infrared. A molecule of oxygen would have a vibration frequency, and energy, smaller by a factor of 4.

Compare the hydrogen vibration and rotation energies with thermal energies at several temperatures:

$$E_{10\ \mathrm{K}} = \tfrac{3}{2}kT = \tfrac{3}{2}(1.38 \times 10^{-23}\ \mathrm{J/K})\ (10\ \mathrm{K}) = 1.3 \times 10^{-3}\ \mathrm{eV}.$$

$$E_{100\ \mathrm{K}} = 1.3 \times 10^{-2}\ \mathrm{eV}.$$

$$E_{1000\ \mathrm{K}} = 1.3 \times 10^{-1}\ \mathrm{eV}.$$

(This $k$ is the Boltzmann constant, not the atomic spring constant.) Compare these threshold energies with the curve of specific heat of hydrogen, as shown in figure 9.4.

Below 10 K, only the translational motion of the molcule can share energy. Above 100 K the rotations can share in the energy. Vibrations don't share completely in the thermal energy until the temperature is close to 1000 K.

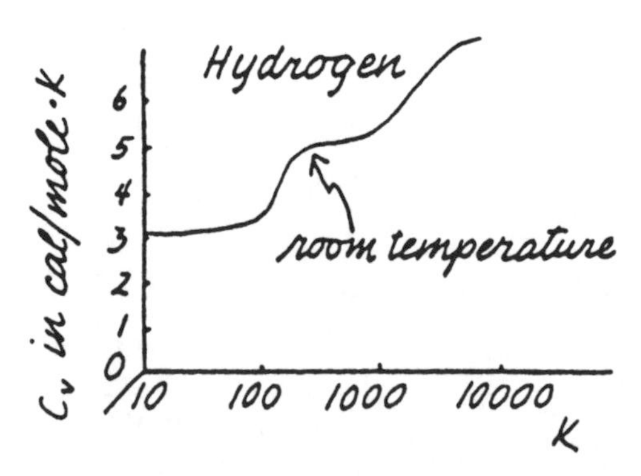


FIGURE 9.4 Specific heat of hydrogen as function of temperature.

Chapter

# 10

# Particles and Quanta

## UNCERTAINTIES OF PHOTONS AND MESONS

Here's a way to defeat the conservation of energy law—apparently. Heisenberg's uncertainty relationships link energy and time interval measurements:

$$\Delta E \cdot \Delta t \geq \hbar = 1 \times 10^{-34}\ \mathrm{J \cdot s},$$

where $\Delta E$ is the uncertainty of the energy of a system that is being measured during a time interval that has an uncertainty of $\Delta t$.

A visible light photon from an atomic transition has a (coherence) length of about 1 meter. The uncertainty in the time of its generation is

$$\Delta t = \frac{1\ \mathrm{m}}{3 \times 10^8\ \mathrm{m/s}} \approx 10^{-8}\ \mathrm{s}.$$

The uncertainty of the photon energy is

$$\Delta E = \frac{1 \times 10^{-34}}{10^{-8}} = 10^{-26}\ \mathrm{J} = 10^{-7}\ \mathrm{eV}.$$

This is the intrinsic line width of an atomic transition that produces a visible photon. The line has a width relative to its own energy of $10^{-7}$, which can just be resolved with high-quality spectroscopy. (There is also a widening caused by the Doppler shift due to thermal motion, but this can be eliminated or avoided.) The electromagnetic field produced by a charged particle can be described in terms of the emission and absorption of virtual photons. They exist long enough to travel out to where they interact, and back again. The further out they go, the smaller their energy must be, giving rise to the $1/r^2$ dependence.

On a different energy scale, a meson, the agent of the nuclear binding force, can be emitted by a nucleon, interact with an adjacent nucleon, and then be reabsorbed, even though it might appear that energy is not conserved in the process. Where does the mass of the meson come from? As long as the fluctuation of energy takes place in a short time interval, satisfying Heisenberg's relationship, the event can take place. The time need only be sufficient for the meson to cross a nucleon diameter since the nuclear force is a short-range force:

$$\Delta t = \frac{10^{-15}\ \text{m}}{3 \times 10^{8}\ \text{m/s}} = \frac{1}{3} \times 10^{-23}\ \text{s}.$$

(This is the nuclear timescale, even as $10^{-8}$ s is the atomic time scale.) The uncertainty, or fluctuation, of energy that can take place in this time is

$$\Delta E = \frac{1 \times 10^{-34}\ \text{J} \cdot \text{s}}{\frac{1}{3} \times 10^{-23}\ \text{s}} = 3 \times 10^{-11}\ \text{J} = 2 \times 10^{8}\ \text{eV} = 200\ \text{MeV}.$$

This is approximately the mass of a $\pi$ meson!

## VALENCE ELECTRONS

### *Velocity of Valence Electron*

The position of the outer electron in an atom cannot be measured closer than about the atomic radius. Applying Heisenberg's uncertainty relationship to an electron in a box:

$$\Delta p \cdot \Delta x \geq \hbar \quad \Rightarrow \quad \Delta p \cdot (10^{-10}\ \mathrm{m}) = (1 \times 10^{-34}\ \mathrm{kg \cdot m^2/s})$$

$$\Delta p \approx p = 1 \times 10^{-24} = m_{\text{electron}} v = (1 \times 10^{-30}\ \mathrm{kg}) v$$

$$v = 10^6\ \mathrm{m/s}\,.$$

Note that this velocity is much less than $c$, and so our nonrelativistic calculation is justified.

### *Kinetic Energy of Valence Electron*

Once again, assume $p \approx \Delta p \approx 1 \times 10^{-24}$

$$E_{\text{kin}} = \frac{p^2}{2m} = \frac{10^{-48}}{2 \times 10^{-30}} = 5 \times 10^{-19}\ \mathrm{J} \approx 3\ \mathrm{eV}$$

### *Wavelength of Photon Emitted by Valence Electron*

As the electron falls into orbit, the wavelength of the emited photon is as follows:

$$E_{\text{pot}} = -k\frac{q}{r}\,, \qquad \frac{mv^2}{r} = k\frac{q}{r^2}$$

$$E_{kin} = \frac{1}{2}mv^2 = \frac{1}{2}k\frac{q}{r}$$

$$E_{\text{total}} = E_{\text{pot}} + E_{\text{kin}} = -k\frac{q}{r} + \frac{1}{2}k\frac{q}{r} = -\frac{1}{2}k\frac{q}{r}.$$

The total energy is negative since the electron is bound. The magnitude of the kinetic energy is equal to the magnitude of the total energy, which equals $5 \times 10^{-19}$ J. The energy of the photon is

$$h\nu = \frac{hc}{\lambda} = \frac{(6 \times 10^{-34})(3 \times 10^8)}{\lambda} = 5 \times 10^{-19}\ \mathrm{J}$$

$$\lambda = 4 \times 10^{-7}\ \mathrm{m}.$$

This is the wavelength of a photon at the blue end of the visible spectrum.

## CAN ELECTRONS RESIDE IN THE NUCLEUS?

Electrons (beta rays) are certainly emitted from nuclei, but are they there to begin with?

Nuclei are composed of neutrons and protons that add in volume as if they were solid balls. Consequently, the volume of a nucleus is proportional to the number of nucleons, which is approximately the atomic "weight," $A$. Thus the nuclear radius is proportional to $A^{\frac{1}{3}}$. A good fit to the data for nuclei radii is

$$r = r_0 \, A^{\frac{1}{3}}, \text{ where } r_o = 1.3 \times 10^{-15} \text{ m.}$$

Consider an atom of aluminum, for which $A^{\frac{1}{3}} = 3$. (The argument is still valid for uranium, where $A^{\frac{1}{3}} = 6.2$.) The radius of the aluminum nucleus is $4 \times 10^{-15}$ m. We know that there are nucleons in the nucleus, but the position of any nucleon is uncertain by a distance equal to the nuclear radius. From Heisenberg's uncertainty principle, we know the momentum of a nucleon must be uncertain by an amount

$$\Delta p = \frac{\hbar}{\Delta r} = \frac{1 \times 10^{-34}}{4 \times 10^{-15}} = 2.5 \times 10^{-20}.$$

Since the momentum, $p$, may be as large as $\Delta p$, the kinetic energy of a nucleon may be as large as

$$E_{\text{kin}} = \frac{(\Delta p)^2}{2m} = \frac{(2.5 \times 10^{-20})^2}{2 \times 1.7 \times 10^{-27}} = 1.8 \times 10^{-13} \text{ J} = 1.1 \text{ MeV.}$$

It takes a bombarding particle with energy of several MeV to dislodge a nucleon from a nucleus. Therefore, a nucleon with a kinetic energy of only 1 MeV can remain bound in a nucleus. Note also that we used the classical relationship between energy and momentum. The nucleon was not in the relativistic regime.

But for an electron with mass very small compared with a nucleon, the relativistic relationship between energy and momentum is

$$E^2 = p^2c^2 + (m_0c^2)^2.$$

Since the mass of the electron is only $1/2$ MeV, we can ignore the rest mass term:

$$E \approx pc = (2.5 \times 10^{-20})(3 \times 10^8) = 7.5 \times 10^{-12}\ \mathrm{J} = 47\ \mathrm{MeV}.$$

An electron with that much kinetic energy could not be bound in a nucleus, and, indeed, the beta rays that are emitted have energies of less than 1 MeV.

## LUCKY MILLIKAN

At first telling, Millikan's oil droplet experiment seems simple. You construct a small parallel plate capacitor in a cavity so that a vertical electric field can be created. At the top leave a small hole in the middle so that you can drip a mist of oil between the plates. Arrange transparent sides on the cavity so that a beam of light can enter in one direction and a telescope can observe along the perpendicular direction. In the telescope you see a small dot of light reflected from each of the oil droplets. They drift downward because of their weight, but many are electrically charged. If you turn on the electric field, the droplets that have positive charge will drift downward even faster and the ones with negative charge will drift upward. If you focus on one droplet and adjust the voltage across the plates, you can balance the forces on that droplet and make it stand still:

$$mg = E\,q \quad \Rightarrow \quad \left(\frac{4}{3}\pi r^3\right)\rho\, g = \frac{V}{\ell}\,q.$$

We assume that the droplet is a sphere and that the oil has density $\rho$. The electric field strength between the capacitor plates is $V/\ell$, where $\ell$ is the spacing between the plates.

Typical values for this equation are

$$\frac{4}{3}\pi(5 \times 10^{-7}\ \text{m})^3\ (850\ \text{kg/m}^3)\ (9.8\ \text{N/kg})$$
$$= \frac{546\ \text{V}}{(0.02\ \text{m})}\ (1.6 \times 10^{-19}\ \text{C}).$$

The apparatus is shown schematically in figure 10.1.

What could be simpler? Millikan found that the charge $q$ was quantized, that is to say, was equal to $ne$, where $n$ is an integer and thus $e$ is the unit of charge. It is the charge on the electron that Millikan measured to an accuracy of better than 1%. The quantization of charge made itself apparent as a quantization of the imposed voltage. In the equation above, for instance, $V$ could equal 273 V, 546 V, or 819 V for $n = 3, 2, 1$.

Now there are two catches. First, how do you measure the radius of an oil drop? Our assumption here was that the diameter of the

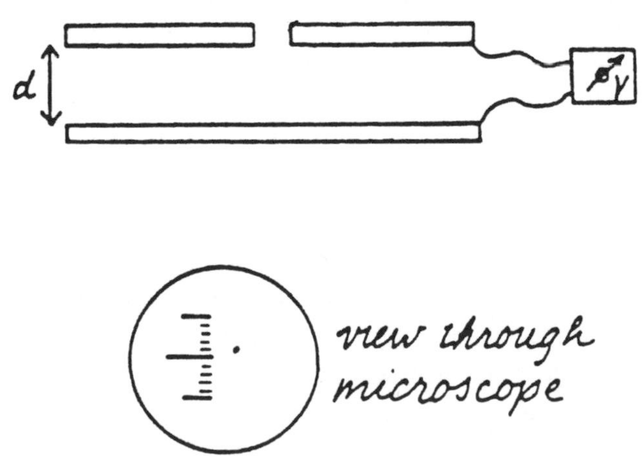


FIGURE 10.1 Schematic drawing of Millikan's oil drop apparatus.

drop was 1 micron. That's a typical size, and nowadays you can get a powder consisting of plastic spheres guaranteed to have a diameter of 1 micron to 1% precision. Millikan, however, and countless generations of physics graduate students since, had to use oil drops and somehow measure the diameter of each. Why not just measure the diameter with the telescope, lining up the edges of the drop with the built-in reticule? But a micron is only twice the wavelength of visible light. The diffraction effects preclude measuring an object with a probe that's about the same size.

To find the mass, Millikan used an equation derived by Stokes. The drag force on a tiny sphere of radius $r$, falling with velocity $v$, through air with viscosity $\eta$ (and the Reynold's number $< 1$) is

$$F_{\text{drag}} = 6\pi\,\eta\,r\,v.$$

When $F_{\text{drag}} = mg$, the droplet falls with terminal velocity, $v$. The most difficult value to measure with precision is the viscosity of air, which is not completely independent of the velocity of the falling object.

The second reason that Millikan's experiment might not work is that the droplets are subject to Brownian motion. Indeed, Brownian motion is often displayed to students using a light cell similar to the one used in the Millikan experiment. The droplets are about the same size as the smoke particles used to demonstrate Brownian motion. The droplets must be in thermal equilibrium with the air molecules. Their average kinetic energy is $\frac{1}{2}mv^2 = \frac{3}{2}kT$, where $k$ is Boltzmann's constant and $T$ is the temperature:

$$v = \sqrt{\frac{3kT}{m}} = \sqrt{\frac{3(1.4 \times 10^{-23})(293)}{\frac{4}{3}\pi(5 \times 10^{-7})^3(850)}} \approx 5\ \text{mm/s}.$$

If the droplets really did dash around at that speed, without bumping into anything, they would never stay long in the field of view of the telescope. They do bump into things, of course—the air molecules.

Their progress is one of slow diffusion. The Brownian motion can be seen as a slight jiggling of the points of light that we identify as droplets, and the Brownian effect makes the drops gradually drift out of focus or out of the light beam.

Millikan was lucky! If the droplets had to be ten times smaller, the Brownian motion would have gotten them. If the droplets had to be ten times larger, the voltages required to provide the balancing E field would have been unattainable.

## NUCLEAR REPULSION

In the Sun, hydrogen is fused through a number of steps into helium, yielding a lot of energy in the process. On Earth we have done the same thing to produce a hydrogen bomb, but have not yet found a way to do it in a gently controlled manner. Why not shoot one proton against another, perhaps in hydrogen gas, and thus produce deuterium in the fusion? Even if the collision problem could be solved, the nuclear interaction would be very small. To change one of the protons into a neutron, a positron and neutrino would have to be emitted, and that involves the *weak interaction.*

Let's calculate how much energy it would require to bring two deuterons close enough together to produce helium from fusion. That reaction is allowed by the *strong interaction.* The radius of a deuteron is about $1.5 \times 10^{-15}$ m. When two deuterons are touching, their center-to-center difference is twice the radius.

The potential energy of two charged particles separated by a distance $r$ is

$$U = k\frac{q_1 q_2}{r}. \text{ This is a potential hill, not a well.}$$

$$U = 9 \times 10^9 \frac{(1.6 \times 10^{-19}\text{ C})^2}{(3 \times 10^{-15}\text{ m})} = 7.7 \times 10^{-14}\text{ J} = 0.48\text{ MeV}.$$

This looks very promising! We can accelerate deuterons to an energy of 0.48 MeV with a table-top machine. Of course, there are problems. Because momentum must be conserved, the bombarding deuteron would have to have twice the penetration energy in order to provide the kinetic energy for the fused pair.

$$mv_o = 2mv_f, \quad v_f = \frac{1}{2}v_o,$$

$$\frac{1}{2}mv_o^2 = \frac{1}{2}(2m)v_f^2 + E_{\text{fusion}} = \frac{1}{2}(2m)\left(\frac{1}{2}v_o\right)^2 + E_{\text{fusion}}$$

$$E_{\text{kin,o}} = E_{\text{kin,f}} + E_{\text{fusion}}$$

$$= \frac{1}{2}E_{\text{kin,o}} + E_{\text{fusion}} \quad \Rightarrow \quad E_{\text{fusion}} = \frac{1}{2}E_{\text{kin,o}}.$$

No need to be discouraged. Deuterons with energy of 1 MeV are still easy to produce with room-sized apparatus. However, deuterons shot into heavy hydrogen gas, or liquid, or solid lose most of their energy in ionizing the target deuterium. Very few deuterons smash head-on into deuterium nuclei.

Why not simply heat up the deuterium until all the ionized deuterons are dashing about and running into one another with high thermal energies? The temperature corresponding to 1 MeV is

$$1 \text{ MeV} = 1.6 \times 10^{-13} \text{ J} = \frac{3}{2}kT$$

$$= \frac{3}{2}(1.4 \times 10^{-23})T \quad \Rightarrow \quad T = 7.6 \times 10^9 \text{ K}.$$

For that temperature, we need a star—or a fission explosion.

The Brookhaven National Laboratory accelerator, RHIC (Relativistic Heavy Ion Collider), can accelerate gold ions in counter-rotating beams. At several places in the giant ring, the beams intersect, producing collisions of gold nuclei with their center of mass at rest in the laboratory frame. Thus, all the kinetic energy can go into collision energy. The Coulomb potential that must be surmounted for nuclear penetration is

$$U = 9 \times 10^9 \frac{(79 \times 1.6 \times 10^{-19}\ \mathrm{C})^2}{(5.8 \times 2.6 \times 10^{-15}\ \mathrm{m})} = 9.6 \times 10^{-11}\ \mathrm{J} = 600\ \mathrm{MeV}.$$

The kinetic energy of each gold nucleus in RHIC is $2 \times 10^4$ GeV. The available energy for collision is twice this. At these energies, well above the Coulomb barrier, the individual nucleons are smashed into their constituent quarks and gluons.

## CYCLOTRON

There's a peculiar relationship between magnetic field strength and the orbital period of a charged particle traveling in a circular path in that field. The period is independent of the radius of the orbit or the energy of the particle. When a particle starts out with a velocity in the plane perpendicular to the magnetic field, the field produces a centripetal force on the particle:

$$\frac{mv^2}{r} = Bev \quad \Rightarrow \quad \frac{2\pi}{T} = 2\pi f = B\frac{e}{m}.$$

Because the period is independent of orbital radius, a driving electric force can be periodic at the cyclotron frequency and accelerate all of the particles regardless of their radial position. Let's put in numbers to find the resonant frequency. A typical proton accelerator would use a magnetic field of 1.5 tesla:

$$f = \frac{(1.5\ \mathrm{T})}{2\pi} \frac{(1.6 \times 10^{-19}\ \mathrm{C})}{(1.7 \times 10^{-27}\ \mathrm{kg})} = 22.5\ \mathrm{MHz}.$$

That's in the radio frequency range with a wavelength of 13 m. The oribtal period is $4.5 \times 10^{-8}$ s.

Note that the resonant frequency does depend on the mass of the particle. The original type of cyclotron could not accelerate protons to energies much higher than 10 MeV. At that energy the proton mass would be relativistically increased by one-tenth of a percent.

Because of the repetitive cycling, however, that difference would be enough to throw the circling proton out of synchronism with the radio-frequency accelerating field.

## SYNCHROTRON—LOSS AND GAIN

There used to be a good argument why electrons could not be accelerated to high energies in circular machines. Charged particles radiate when they are accelerated. The centripetal acceleration in a circular machine would cause the electron energy to drain away. The problem is much more acute for electrons than for protons or heavier particles for the following reason. The magnetic force of the bending magnet provides the centripetal force needed for circular motion:

$$Bqv = \frac{\gamma m v^2}{r} = \gamma m a_c \quad \Rightarrow \quad a_c = B\frac{q}{\gamma m}v$$

$$\left(\gamma \text{ is the relativistic factor } = \frac{1}{\sqrt{1 - v^2/c^2}}\right)$$

For either protons or electrons, the velocity in the large accelerators is close to the speed of light, and so is essentially constant. The centripetal acceleration is then inversely proportional to the mass of the particle.

Physicists realized in the 1960s that this flaw in electron acceleration could be turned into a virtue by making use of the emitted radiation. At high relativistic velocities (high $\gamma$), the light comes off in the forward direction. At high enough electron energy, the radiation is in the X-ray region. Here was a source of very high intensity X-rays with many features of timing and polarization under control of the experimenter.

The power emitted (in a foreward cone) by a charged particle accelerated in a direction perpendicular to its velocity (see D. J.

Griffiths, *Introduction to Electrodynamics* [Prentice Hall 1989], p. 432) is

$$P = \frac{1}{4\pi\varepsilon_0}\frac{2}{3}\frac{q^2a^2}{c^3}\gamma^4.$$

The National Synchrotron Light Source at Brookhaven National Laboratory has the following characteristics:

electron energy, 2.6 GeV

average radius, $r$, 27 m

$$\text{electron orbital period, } \frac{170 \text{ m}}{3 \times 10^8 \text{ m/s}} = 5.7 \times 10^{-7} \text{ s}$$

$$\gamma = \frac{1}{\sqrt{1 - v^2/c^2}} = \frac{\gamma m}{m} = \frac{2.6 \times 10^9 \text{ eV}}{5.1 \times 10^5 \text{ eV}} = 5.2 \times 10^3$$

$$v = c \qquad a = \mathrm{v}^1/r = c^2/r$$

$$P = \frac{1}{6\pi(8.9 \times 10^{-12})}(1.6 \times 10^{-19})^2\frac{(3 \times 10^8)}{(27)^2}(5.2 \times 10^3)^4$$
$$= 4.6 \times 10^{-8} \text{ J/s} = 2.9 \times 10^{11} \text{ eV/s}.$$

The energy lost per revolution is

$$\Delta E = P \times T = (2.9 \times 10^{11} \text{ eV/s})(5.7 \times 10^{-7} \text{ s/turn})$$
$$= 1.6 \times 10^5 \text{ eV/turn}.$$

In order to maintain its orbit, this much energy per turn is fed into the electron at several acceleration stations around the ring. The beam bending is not done continuously as in our approximation, but takes place at a number of bending magnets along the orbit. The beam is also sent through magnetic "wigglers" that produce large periodic accelerations yielding high-intensity X-rays.

# QUANTIZED MOLECULAR LEVELS

The energy levels of atoms bound in a solid are separated by energies $h\nu$, where $h$ is Planck's constant and $\nu$ is the atomic vibration frequency. Each atom can be pictured as existing in a (three-dimensional) potential energy well. The greater its vibration energy, the higher it is in the well. However, the only energies allowed must satisfy the quantum condition, $E = (n + \frac{1}{2})h\,\nu$, where $n$ is an integer.

For uranium, the fundamental vibration frequency is

$$\nu = \frac{1}{2\pi}\sqrt{\frac{k}{m}} = \frac{1}{2\pi}\sqrt{\frac{50}{(238 \times 1.7 \times 10^{-27})}} = 1.8 \times 10^{12}\ \text{Hz}.$$

(For the value of the spring constant, $k$, see "Atomic Spring Constant" in chapter 9.) This corresponds to energy level separations of

$$h\,\nu = 1.2 \times 10^{-21}\ \text{J} = 0.0070\ \text{eV} = \frac{1}{140}\ \text{eV}.$$

In a potential well that is about 3 eV deep, there are about four hundred such levels. Room-temperature thermal energy of a particle is about $^1/_{25}$ eV. (See "Thermal Expansion" in chapter 4.) This is much larger than the quantized energy levels of uranium, and so the specific heat of uranium has the classical DuLong-Petit value until the temperature falls to about 40 K.

For diamond, however, the depth of the potential energy well for the carbon atoms is about 7 eV. This corresponds to a maximum binding force of

$$F = \frac{\Delta U}{\Delta r} = \frac{7\ \text{eV} \times 1.6 \times 10^{-19}\ \text{J/eV}}{1 \times 10^{-10}\ \text{m}} = 11 \times 10^{9}\ \text{N}.$$

The spring constant is $k = \frac{F}{r} = \frac{11\times10^{-9}\ \text{N}}{1\times10^{-10}\ \text{m}} = 110\ \text{N/m}$.

The bonds between the carbon atoms in diamond are extremely strong. Because of the small mass of the carbon atom, the frequency

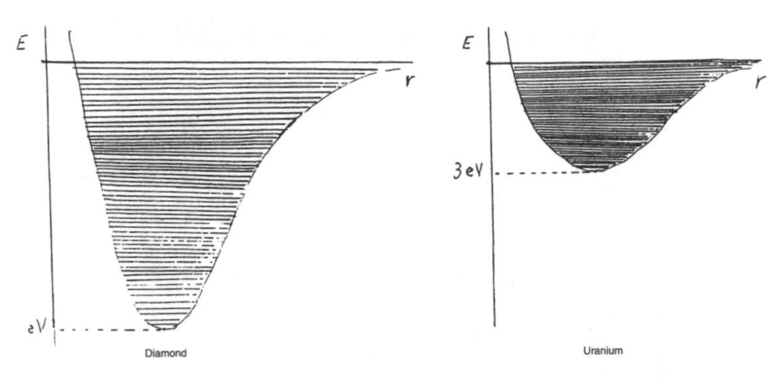


FIGURE 10.2 Spacing of energy levels in diamond and uranium.

of vibration in the crystal is very high, and the equivalent energy levels are large.

$$\nu = \frac{1}{2\pi}\sqrt{\frac{110}{(12)\times(1.7\times10^{-27})}} = 1.2\times10^{13}$$

$$h\,\nu = (6.6\times10^{-34})(1.2\times10^{13}) = 7.7\times10^{-21}\text{ J}$$

$$= 0.048\text{ eV} = \frac{1}{21}\text{ eV}.$$

At room temperature or lower, diamond is inefficient at absorbing thermal energy because a collision with another particle usually cannot provide enough energy for the diamond atom to jump to a higher level. Consequently, the specific heat of diamond climbs slowly as a function of temperature and does not reach full value until $T > 1000$ K.

See figure 10.2 for two energy-level diagrams, one for diamond and one for uranium. The depth of the diamond well is 7 eV. It has about 150 energy levels, each separated by 1/21 eV. The uranium well is shallower, corresponding to 3 eV binding energy of the atom in the solid. There are about four hundred energy levels, each separated by 1/140 eV.

# MOLECULAR ROTATION ENERGY

The smallest unit of angular momentum change is the reduced Planck's constant, $\hbar$. Let's calculate the energy required when a hydrogen molecule increases its angular momentum by one unit. The two protons are spinning around each other at a radial distance for each of $0.37 \times 10^{-10}$ m.

$$\Delta(\text{angular momentum}) = \Delta(I\,\omega) = \hbar$$

$$\text{Moment of inertia of hydrogen molecule } = I = 2m_{\text{p}}r^2$$

$$\text{Rotational kinetic energy } = E_{\text{r}} = \frac{1}{2} I\,\omega^2 = \frac{1}{2}\frac{(I\,\omega)^2}{I} = \frac{1}{2}\frac{\hbar^2}{I}$$

$$E_r = \frac{1}{2}\frac{(1 \times 10^{-34})^2}{2(1.7 \times 10^{-27})(0.37 \times 10^{-10})^2}$$

$$= 1.5 \times 10^{-21}\ \text{J} \approx 1 \times 10^{-2}\ \text{eV}.$$

What realm of the electromagnetic spectrum is this? Microwave radiation goes down to $\lambda$ of about 1 cm. The photon energy for this wavelength is $h\,\nu = \frac{hc}{\lambda} = \frac{(6.6\times10^{-34})(3\times10^{8})}{1\times10^{-2}} = 2.0 \times 10^{-23}\ \text{J} = 1.2 \times 10^{-4}\ \text{eV}$. Apparently, microwave energy is too small to excite rotation of the hydrogen molecule. However, note that in the rotational energy formula, the moment of inertia is in the denominator. For heavier and larger molecules, the rotational kinetic energy steps are small enough to be excited, and thus studied, by microwaves.